CANADIAN COUNTERCULTURES

COUNTERCULTURES

AND THE

ENVIRONMENT

Canadian History and Environment Series
Alan MacEachern, Series Editor
ISSN 1925-3702 (Print) ISSN 1925-3710 (Online)

The Canadian History & Environment series of edited collections brings together scholars from across the academy and beyond to explore the relationships between people and nature in Canada's past. Published simultaneously in print and open-access form, the series then communicates that scholarship to the world.

ALAN MACEACHERN, FOUNDING DIRECTOR
NiCHE: Network in Canadian History & Environment
Nouvelle initiative canadienne en histoire de l'environnement
http://niche-canada.org

UNIVERSITY OF CALGARY
Press

EDITED BY **Colin M. Coates**

CANADIAN COUNTERCULTURES AND THE ENVIRONMENT

Canadian History and Environment Series
ISSN 1925-3702 (Print) ISSN 1925-3710 (Online)

University of Calgary Press
2500 University Drive NW
Calgary, Alberta
Canada T2N 1N4
www.uofcpress.com

LIBRARY AND ARCHIVES CANADA CATALOGUING IN PUBLICATION

Canadian countercultures and the environment / edited by Colin Coates.

(Canadian history and environment series, 1925-3702 ; no. 4)
Includes bibliographical references and index.
Issued in print and electronic formats.
ISBN 978-1-55238-814-3 (paperback).–ISBN 978-1-55238-816-7 (pdf).–
ISBN 978-1-55238-817-4 (epub).–ISBN 978-1-55238-818-1 (mobi).–
ISBN 978-1-55238-815-0 (open access pdf)

1. Environmentalism–Canada–History–20th century. 2. Counterculture–
Canada–History–20th century. 3. Canada–Environmental conditions–History–
20th century. I. Coates, Colin MacMillan, 1960-, editor II. Series: Canadian history
and environment series ; 4

GE199.C3C33 2015   333.72097109'047   C2015-907613-7

                                                   C2015-907614-5

The University of Calgary Press acknowledges the support of the Government of Alberta through the Alberta Media Fund for our publications. We acknowledge the financial support of the Government of Canada through the Canada Book Fund for our publishing activities. We acknowledge the financial support of the Canada Council for the Arts for our publishing program.

Cover photo: MAB group die-in, Montreal, 1976. Source: MAB Archives.
Editing by Alison Jacques
Cover design, page design, and typesetting by Melina Cusano

TABLE OF CONTENTS

ACKNOWLEDGMENTS

Fifty years after the counterculture began to have an impact on Canada, this collection revisits the way that various groups who contested mainstream norms in the 1960s, 1970s, and 1980s approached the environment. In 2011, most of the authors in this collection met at an initial workshop, hosted by the Heron Rocks Friendship Centre Society on Hornby Island, British Columbia. Hornby Island, like other Gulf Islands and indeed many other rural parts of British Columbia, experienced the direct impact of the counterculture. On Hornby Island, many people who came to the islands at that time stayed, and they remain key figures in the economy and politics of the island.

The Heron Rocks Friendship Centre Society (heronrocks.ca) is a nonprofit society dedicated to maintaining the vision of local activists Hilary and Harrison Brown and encouraging community and sustainable stewardship of the land. The society provided us space in their exquisite setting under the oak tree and offered us meals. The society's executive, along with its membership, welcomed us with grace and generosity. Rudy Rogalsky played a particularly key role as liaison with the society. Some members attended the workshop, as did other interested people from Hornby and Denman islands, and they shared their views on our discussions. Jan Bevan provided particularly poignant reflections. Margaret Sinclair prepared a photo exhibit on the counterculture period on the island and displayed it at the Hornby Island Co-op. We would like to thank the people on Hornby Island who provided such beautiful accommodations for our early July meeting.

We would also like to thank the external reviewers of the University of Calgary Press for their helpful and constructive comments. This project, like so many others, stems from the Social Sciences and Humanities Research Council–funded Network in Canadian History & Environment/Nouvelle initiative canadienne en histoire de l'environnement. Alan MacEachern, the founding director of NiCHE, was a participant at our workshop, and he played an essential role both in ensuring that this project take place and in encouraging it along the way. On behalf of the contributors to this volume, I would also like to thank Mary-Ellen Kelm, Kathy Mezei, Bob Anderson, Lauren Wheeler, Peter Evans, and Terry Simmons, colleagues who also attended the workshop and provided thoughts on this project. Peter Enman and Melina Cusano at the Press assisted us through the process, and Alison Jacques contributed her exemplary copyediting skills. Finally, many participants in the countercultural activities that are covered in this book shared their memories, interpretations, and documents, and other scholars kindly provided research notes and documents that they had uncovered in the course of their own projects. The subject of countercultures and the environment clearly inspires generosity.

CONTRIBUTORS

MATT CAVERS studied historical geography at the University of British Columbia and Queen's University. He now works as a brewer at a craft brewery and hop farm on British Columbia's Sunshine Coast.

COLIN M. COATES is director of the Robarts Centre for Canadian Studies, and he teaches environmental history and Canadian studies at York University. He is past president of the Canadian Studies Network–Réseau d'études canadiennes.

MEGAN J. DAVIES is an associate professor at York University and a BC historian with research interests in madness, marginal and alternative health practices, old age, rural medicine, and social welfare. Her recent projects include *After the Asylum,* a research site about the history of deinstitutionalization in Canada, and the documentary film *The Inmates Are Running the Asylum: Stories from MPA.*

NANCY JANOVICEK teaches Canadian history and gender history at the University of Calgary. She is the author of *No Place to Go: Local Histories of the Battered Women's Shelter Movement* (UBC Press) and co-editor of *Feminist History in Canada: New Essays on Women, Gender, Work, and Nation* (UBC Press).

ALAN MACEACHERN teaches history at the University of Western Ontario and is the founding director of NiCHE: Network in Canadian History & Environment. He has written extensively on Prince Edward Island in the 1970s, including *The Institute of Man and Resources: An Environmental Fable* (2003).

DAVID NEUFELD, an environmental historian living and working in the Yukon Territory, studies the intersection of knowledge and practice in settler and First Nations approaches to their shared subarctic boreal homeland. His reflexive research approach is grounded in thirty years as a community-based cultural researcher for Parks Canada.

RYAN O'CONNOR Ryan O'Connor is the author of *The First Green Wave: Pollution Probe and the Origins of Environmental Activism in Ontario* (UBC Press, 2015). An Associate Member of the Centre for Environment, Heritage and Policy at the University of Stirling (Scotland), he maintains a research blog at www.ryanoconnor.ca.

KATHLEEN RODGERS teaches sociology at the University of Ottawa. She is the author of *Welcome to Resisterville: American Dissidents in British Columbia* (UBC Press) and co-editor of *Protest and Politics: The Promise of Social Movement Societies* (UBC Press).

DANIEL ROSS is a PhD candidate in history at York University. He studies urban, political, and environmental history and is currently completing a dissertation on the politics of development and public space in downtown Toronto from the 1940s onwards.

HENRY TRIM is a Social Sciences and Humanities Research Council of Canada fellow at the University of California, Santa Barbara, where he studies forecasting and sustainability in North America. He has written about science and American counterculture, environmental politics, and Canadian energy policy.

SHARON WEAVER completed her PhD at the University of Guelph in 2013. Her dissertation, "Making Place on the Canadian Periphery: Back-to-the-Land on the Gulf Islands and Cape Breton," is a comparative study of back-to-the-landers in Cape Breton and on the Gulf Islands of British Columbia. She is an independent scholar who lives in Fredericton, New Brunswick.

Canadian Countercultures and their Environments, 1960s–1980s

Colin M. Coates

"Happiness," declared twenty-three-year-old hippie John Douglas to a *Toronto Star* reporter in 1967, "is hauling water from the stream."[1] For the former Torontonian, then living on a farm in the Madawaska Highlands in northern Ontario, this communion with nature was a novelty. It is not inconceivable that Douglas's parents and, even more likely, his grandparents spent part of their days fetching water and carrying it into their houses. Whatever they thought about their living circumstances, they were likely more inured to and less ecstatic about the task. But for the young Douglas, the physical chore involved a spiritual component, illustrating the links that many people who chose a counterculture lifestyle consciously made to the environment. A direct experience of nature represented a moral choice for many during this period of cultural upheaval associated with the counterculture from the 1960s to the 1980s.

Covering a range of case studies from the Yukon to Atlantic Canada, this book explores the ways in which Canadians who identified with rural and urban countercultures during the 1960s, 1970s, and 1980s engaged with environmental issues. This awareness covered a broad range of areas, from celebrations of the human body to concerns about environmental degradation. Throughout Canada, groups of young people established alternative communities and consciously embraced new practices. Their choices led them to connect with environmental issues in innovative and committed ways.

The book is divided into two sections. The first section explores examples of environmental activism and focuses on innovative local organizing and advocacy. The second section examines countercultural life choices and the environmental perspectives these entailed. Technological options, relations with the state, and encounters with hostile and curious local populations all held particular implications for people espousing alternative lifestyles.

This chapter presents the broad contours of countercultural environmentalism across Canada and introduces the key themes of this collection of essays. In exploring the broad connections between the Canadian counterculture and environmental issues, it makes the point that this truly was a pan-Canadian phenomenon, including Francophones and Anglophones from coast to coast to coast. At the same time, this was an international movement, and the influx of American men and women, many of whom were critical of the Vietnam War, reinforced the oppositional stances of Canadian youth. Many were inspired by utopian sentiments, and they moved to rural communes to live out their ideals, in places where they engaged of necessity with the natural environment in a very direct way. Scholars who deal with utopian societies tend to focus on the ultimate failures. In contrast, this book insists on the legacies of the Canadian counterculture. Much of the countercultural critique of contemporary attitudes to the environment has become mainstream today.

Of course, not all back-to-the-landers chose to live in communes. The majority homesteaded. Nonetheless, as this collection illustrates, commune-dwellers and non-commune-dwellers shared many utopian and environmental perspectives and experiences. This chapter draws on my research on Canadian utopian settlements, and therefore it accentuates the experiences of counterculture communes.

COUNTERCULTURES

Drawing from earlier generations of youthful disaffection, people across North America and throughout the Western world in the 1960s and 1970s engaged in activities associated with the "counter-culture." Three key contexts in which the counterculture developed were the Vietnam War, the baby boom demographic bulge, and the connected rise of 1960s youth culture. The Vietnam conflict heightened both anxiety about Cold War military confrontations and fear among many young American men of being drafted to fight in a distant and unpopular war. American men and women took refuge in Canada, whether from the military draft or simply from the politics of their country. The decision could reflect more of a personal decision to escape the troubles of the period: writer Mark Vonnegut left the East Coast of the United States in order to acquire land in British Columbia, positing, "I think the Kennedys, Martin Luther King, and war and assorted other goodies had so badly blown everybody's mind that sending the children naked into the woods to build a new society seemed worth a try."[2] Americans and Canadians moved to relatively remote areas, searching for affordable land. National identity was not irrelevant, but young Americans and Canadians shared a dislike of American military policies and both participated fully in a broad Western international youth culture.[3]

Often associated with "hippies," the term "counterculture" flattens many differences. As Peter Braunstein and Michael William Doyle point out, the concept encompassed a wide variety of attitudes, practices, beliefs, and styles.[4] One of the key Canadian activists of the period, Greenpeace founder Bob Hunter, sums up the variety of people in Vancouver, British Columbia, who supported countercultural environmentalism:

> We had the biggest concentration of tree-huggers, radicalized students, garbage-dump stoppers, shit-disturbing unionists, freeway fighters, pot smokers and growers, aging Trotskyites, condo killers, farmland savers, fish

preservationists, animal rights activists, back-to-the-landers, vegetarians, nudists, Buddhists, and anti-spraying, anti-pollution marchers and picketers in the country, per capita, in the world.[5]

Greenpeace was itself one of the major Canadian contributions to environmentalism in the late twentieth century; its story has been well covered by Frank Zelko.[6] But beyond this large, soon-to-be international organization, many people organized on the local level to make innovative choices concerning the environment. Moving "back to the land" reflected one expression of the counterculture, and the destination required a deep engagement with ecological realities. However, as this collection shows, people who remained in urban centres also contributed to changing perspectives on environmental issues. Many of the people whose stories are recounted in this collection rejected an affluent and consumer-oriented urban culture and chose a different, usually rural, path. Political scientist Judith I. McKenzie provides a helpful definition of "counterculture": a "deliberate attempt to live according to norms that are different from, and to some extent contradictory to, those institutionally enforced by society, and oppose traditional institutions on the basis of alternative principles and beliefs."[7] It is significant that many of the people at the time adopted the term "counterculture" to describe themselves and their choices. But historian Stuart Henderson adds an insightful coda to definitions of counterculture: "In his or her rejection of [the] dominant culture, the hippie is in fact operating within, *not without*, the same culture. . . ."[8] Whether urban or rural, counterculturalists in Canada challenged societal norms by choosing to live differently, often in communal arrangements.

The prosperity of the 1950s and 1960s and the demographic bulge of children born after 1945 had created rising expectations and enhanced a youth culture that was rapidly commercialized, but which nonetheless revelled in oppositional perspectives.[9] Youth culture took many forms in the decades that followed. Most youth did not participate meaningfully in the counterculture, though they may on

COLIN M. COATES

occasion have participated in some of its apparently defining characteristics, such as enjoying the music of the period and smoking marijuana or taking other hallucinogenic drugs.[10] This book focuses on those who determinedly attempted to create new social norms.

THE UTOPIAN IMPULSE

John Douglas's counterculture generation was not the first to locate their vision of utopia in the embrace of nature and rural labour and the rejection of the amenities of urban life and consumerism. Throughout Canadian history, utopian dreamers have located their perfectible worlds primarily in the countryside, and therefore one key feature of Canadian utopianism—much like its American counterpart—is its connection to an agrarian, "natural" world. Inspired in part by utopian thinkers, such as nineteenth-century writer Henry David Thoreau, or by twentieth-century nature writers, such as Aldo Leopold in the United States or Grey Owl in Canada, young people in the late 1960s and 1970s streamed into marginal areas throughout North America, away from the cities in which they had been raised. Their preferences had a practical side, as land prices were much lower in the countryside than in urban areas, and there were particularly good deals on lands where agriculture represented a marginal, declining activity. To achieve a utopian society, groups set themselves outside of larger centres and away from consumption-oriented mainstream Canadian society.

For some individuals, Canada offered isolated regions far from the tribulations of urban life. New England professor Feenie Ziner's son Ben escaped to a remote forested island off the West Coast. When she went looking for him in the 1970s, she believed—as he likely had when he arrived there—that she was "flying over the last and final untamed wilderness in North America."[11] Writer Mark Vonnegut ended up in a corner of the Sunshine Coast, not far from Ben's island: "This was virgin frontier, unspoiled except for ugly scars left by loggers here and there. Man was here but not many of 'em and he was certainly not master."[12] Some chose their lands specifically in order to be at some distance from state authorities. The participants in a commune near Powell River, BC, spoke wistfully of the "freedom of the country" to a CBC reporter in 1969 who promised not to reveal specifically where they were located.[13] They had reason to be circumspect. Not only could the sudden arrival of enthusiasts overburden a commune's resources, government officials sometimes were very dubious about their efforts. As one of the early scholars of the movement, geographer Terry Simmons found building inspectors who had the job of enforcing local housing regulations could make life very difficult for commune-dwellers.[14]

While isolation was a tremendous draw for a number of political and practical reasons, Ben Ziner's case was more extreme than some. Most back-to-the-landers located in previously settled areas, places where they could grow at least some of their own food. This reflected a political choice addressing fears that global annihilation was at hand: "Time is rapidly running out for Mother Earth. In order to save her we must get our shit togeather [sic] and begin building agricultural communes . . . the base [sic] of the revolution," declared the Marxist-Leninist Ochiltree Commune, near Williams Lake in the interior of British Columbia.[15] Ochiltree was one of the most intensely political communes of the period, but many people elsewhere shared a belief that the political and ecological environment in which they lived was about to explode. Americans Barry and Sally Lamare relocated to New Denver, in southeastern BC, in the mid-1970s because of the apparent security it offered in the case of nuclear war: "It was over fifteen hundred feet in altitude, you see, so it was above radiation levels. You could grow vegetables and survive."[16] Such apocalyptic fears would ultimately serve to weaken the back-to-the-land movement. Historian Michael Egan points out that, when the jeremiads failed to translate into reality with the speed predicted, environmentalist messages lost much of their impact.[17]

Nonetheless, in the short run, self-sufficiency seemed to offer the solution to social instability and ecological fears. In the Bas-St-Laurent region of Quebec, three men and one woman established La Commune de la Plaine in the spring of 1972, based on shared property and a rigorous egalitarianism. They wished to create "the most wide-ranging self-sufficiency possible." Like their counterparts in other parts of the country, their choices involved a spiritual reawakening. As Marc Corbeil, one of the participants, reflected in an academic study some years later, "It was a search for a healthy lifestyle, in contact with nature, for us collectively and individually, where work would regenerate us, and bodily and spiritual pleasures would have their place." The commune survived until 1985, and Corbeil estimated that about one hundred people passed through it during its time.[18]

Even an apparent exception to the "back-to-the-land" ethos provides confirmation of the healing propensities of rural life. Therafields was a large therapeutic commune based in the Annex area of downtown Toronto. One long-term member proposes that it was "arguably the largest secular '60s commune in North America," with about nine hundred adherents in its heyday.[19] Houses along Walmer Road provided the urban residences for the people involved, but many of the key therapeutic sessions took place on the Therafields farm the group owned in Mono Mills, near Orangeville, and from which the community took its name. While on the farm, participants engaged in hard labour, often divided along gendered lines, while spending other times in encounter sessions. For some members, the farming labour seemed more significant than the psychological benefits, even if they resented the hard work. The physicality of the work was conceived as improving the mental health of the individual. In an article explaining the philosophy of the group, the leaders of Therafields juxtaposed their belief that "Society as it has evolved is a robot beyond control"[20] against organic and biological metaphors that show how the group helped individuals overcome the issues they faced. Mind and body were well served by the encounter with nature, even if one had to leave Toronto temporarily to experience it. In August 1978, the group held a "Therafields Country Fair" on their rural site, where they sold organic produce and crafts.[21] Even the most urban commune needed a rural retreat.

Whether the young men and women taking part in the counter-culture were looking for a refuge or a spiritual nirvana, their engagement with their location and their choice of economic activity forced them to confront environmental issues. Such concerns had indeed begun to achieve greater prominence in the 1960s and 1970s, but not exclusively because of the counterculture. Yet it is interesting that a number of observers, including key contemporary figures, point to environmental consciousness as being one of the principal legacies of the counterculture.[22]

ENVIRONMENTALISM

Late-twentieth-century environmentalism has many origin stories—but, normally, it is not closely associated with the organized youth movements of the 1960s. Historians have argued that environmentalism was not a key theme of New Left politics in the United States in the 1960s. The Port Huron Statement of the Students for a Democratic Society made only a brief reference to environmental issues, linking economic growth with ecological problems:

> We cannot measure national spirit by the Dow Jones Average, nor national achievement by the Gross National Product. For the Gross National Product includes air pollution. . . . The Gross National Product includes the destruction of the redwoods and the death of Lake Superior.[23]

Such concerns were fairly mainstream in the 1960s. Rachel Carson's *Silent Spring* had inspired a great deal of the period's environmental consciousness, often focused around pollution and reaching a broad swath of the North American public. Many middle-class, suburban mothers played key activist roles in supporting environmental protection and improvement measures. They worked alongside government and social leaders such as Lady Bird Johnson, wife of American President Lyndon Johnson.[24]

Historian Keith M. Woodhouse argues that a sudden shift occurred after 1969, leading to the first Earth Day in 1970. This event, sponsored by Republican Senator Gaylord Nelson and supported by President Richard Nixon's government, demonstrated how environmental concerns could be seen as liberal rather than radical issues.[25] Contemporaries advanced cynical interpretations of this embrace of environmentalism. Speaking before the Men's Canadian Club of Toronto in 1970, geographer F. Kenneth Hare evaluated the US government's sudden focus on pollution issues thus: "it is convenient for central governments to have an issue that doesn't really divide the electors, that doesn't antagonize the campuses, and that so often doesn't involve any concrete action."[26] In Canada in the 1960s, debates over environmental issues tended to focus around issues of access to wilderness park–like areas.[27] Refracted through the lens of leisure, ecological issues became part of the public agenda.

Perhaps because of its broad appeal, environmentalism quickly entered into popular culture. These are only a few striking examples: American musician Marvin Gaye may have penned one of the best-known environmentalist anthems, "Mercy Mercy Me (The Ecology)" in 1971, but he was preceded by Saskatchewan-born Joni Mitchell's "Big Yellow Taxi" in 1970, a critique of excessive urban development. The 1975 album of the Quebec folk group Les Séguin, "Récolte des Rêves," provided similar, nostalgic celebrations of agrarian lifestyles. Many other musicians adopted ecological themes.

Concerns for the environment may of course take many different forms of expression, ranging from the designation of new park areas, to struggles against pollution, to changing the way one grows food. The archetypal countercultural environmental group of this period, Greenpeace, had its roots in Vancouver's Kitsilano neighbourhood, where it had organized to oppose testing of nuclear bombs on the Aleutian Islands of Alaska. Some of the key figures in the organization took inspiration from oppositional attitudes, drawing on Quakerism along with New Left and peace movement perspectives as well as Marshall McLuhan's communication theories.[28]

In contrast to the worldwide organization that Greenpeace became, the countercultural groups discussed in this book tended to focus on more grassroots local issues, though many participants may have agreed with the founders of La Commune de la Plaine that they were involved in revolution. Certainly the Ochiltree commune in BC did. In fact, as Ryan O'Connor points out, the recycling efforts begun on a small-scale basis in Toronto in the 1970s have become very large worldwide businesses indeed. At the same time, the Ark experiment in sustainable living on Prince Edward Island, which Henry Trim examines, failed to have the broad impact its founders had desired.

Commune-dwellers' beliefs in "voluntary simplicity" and self-sufficiency encouraged and facilitated the adoption of environmental approaches. Having accepted a less materialist lifestyle, labour was consequently fairly cheap. Many communes adopted organic techniques; this choice saved money on the costs of chemical fertilizers and herbicides, and it provided even more work for the people living on the farms. Local commune-dwellers read their copies of the *Whole Earth Catalog* and other works such as Helen and Scott Nearing's *Living the Good Life: How to Live Sanely and Simply in a Troubled World* (1954). A Canadian Council on Social Development survey of communes in Ontario, Quebec, and the Maritimes found that farm communes "were predominantly interested in agricultural subsistence with their main objectives being to farm organically, to have the land meet as many needs as possible, and to make the commune independent and self-supporting."[29] Taking inspiration from the *Whole Earth Catalog* and using a sumac branch as a maple syrup tap, back-to-the-lander Mark Frutkin reminisced about his choice: "I was enamoured of the old ways because they used what was in the environment. For me it was a statement about self-sufficiency."[30]

Choosing self-sufficiency often entailed opting for a fairly marginal economic existence. While many of the youth had their advanced education to fall back on—and of course they knew that—for the time that they lived on the communes, they accepted a different and unfamiliar lifestyle, and for most, it was not an easy one. For instance, some had to wrestle with practical husbandry issues for which they were not prepared. Members of La Commune des plateaux de l'Anse-Saint-Jean, in Quebec's Saguenay region, found it necessary to keep their animals inside on the ground floor of their dwelling during their first winter, while they lived on the top floor.[31] Moving back to the land required direct confrontation with agrarian realities and an environmental consciousness.

ISSUES

Perhaps one of the key ways in which countercultural environmentalism differed from other forms was its emphasis on the body. As recalled by theologian Gregory Baum, who maintained links to Therafields in the early years, the commune's work therapy was inspired by Reichian psychiatry: "The body was taken seriously."[32] People who had made the choice to join the counterculture willingly distinguished themselves from their urban counterparts. They rejected some urban, middle-class niceties, and so chose long hair for both men and women, refused to shave, and practiced public nudity. One member of La Commune des plateaux de l'Anse-Saint-Jean remembered how they differentiated themselves visibly from other locals through their dress and hair.[33] Gardening in the nude did not likely impinge upon neighbours, but bathing without clothes at the beach tended to annoy other members of the community, as was the case on Denman Island in the 1970s. Des Kennedy remembered that nude swimming became a "kind of flash point for a lot of people."[34] Public nudity fed into assumptions of looser sexual norms, which were becoming more prevalent far beyond the counterculture.[35] In fact, members of La Commune des plateaux de l'Anse-Saint-Jean, as well as many others, complained that the perception of wanton sexuality that was attached to many commune-dwellers did not in fact accurately reflect their more moderate lifestyle.[36] Mark Frutkin recalls the lack of debauchery on his commune in the Gatineau region of Quebec:

> Everyone wanted to partner up as soon as possible, although there was almost no sharing of partners and no attempts

at group marriage at the Farm. We must have been the straightest, dullest commune on the face of the planet if the articles in *Life* and *Time* were to be believed.

Nonetheless, the commune-dwellers also practiced public nudity at a nearby lake and in the group saunas. But, Frutkin points out, the prevalence of cold and insects restricted nudity to about two months of the year.[37]

Opting for nudity reflected the desire to reduce the distance between the human body and the environment, an enhancement of authenticity. Following the same logic, many women celebrated the natural process of birth, attempting to reclaim knowledge that in Canada the medical profession had monopolized in the twentieth century. Childbirth had become a medicalized and hospitalized procedure. As Megan Davies shows in discussing underground midwifery in southeastern British Columbia, activists in the 1970s and 1980s fostered the growth of a cadre of trained, but non-professional, midwives, fully engaged with local communities.

As other chapters illustrate, countercultural youth often harboured a suspicion of local development and its potential effects on healthy bodies. As Nancy Janovicek shows in this volume, local counterculture settlers opposed large-scale logging in the Kootenays, in southeastern British Columbia, pointing out how little of the profit from the industry remained in the area. In a complementary chapter, Kathleen Rodgers explores American influences on environmental protest in the Kootenays. With their goals of self-sufficiency, counterculture youth demonstrated an anti-consumerist bias in much of what they did. Daniel Ross shows how cycling activists in Montreal decried the overuse of the car, a message that took hold in part because of the shock of the oil crisis of the early 1970s. In contrast, as Ryan O'Connor argues, recycling advocates in Toronto achieved their greatest success not in reducing consumption, but rather in dealing with the effects of consumerism in a novel way.

In some cases, back-to-the-landers aimed at a highly simplified lifestyle, rejecting modern conveniences. In Carleton County, New

Brunswick, a group of Americans revelled in their marginal and isolated farmstead: "In an electronic, thermostatically controlled world it is all too easy to let insensitivity dull all the sense of feelings. I suppose what we are mostly trying to do here is give these kids a chance to react to their environment and to become more sensitive to living and to the land."[38] Marc Corbeil recalled nostalgically how, at La Commune de la Plaine, "the commune-dwellers had the erroneous impression that old means of production were less complicated."[39]

But equally typically, counterculture youth embraced what they considered appropriate technologies. *Little House on the Prairie*–type technologies still required advanced understanding and skill. Many sought to integrate newer technologies with an aim to self-sufficiency, sometimes taking inspiration from the *Whole Earth Catalog*, which provided scientific models to assist in living off the grid.[40] Quebec readers had their own version of this publication in *Le Répertoire québécois des outils planétaires*. As Henry Trim argues, the Ark experiment on Prince Edward Island grew out of concerns in the 1960s and particularly the early 1970s with spiralling energy costs and rural decline. In this case, the founders tried to develop a sophisticated technology to address issues of self-sufficiency and provide a model that could be replicated elsewhere. For many people drawn to the counterculture, as Walter Isaacson shows in the case of Californians associated with the development of the personal computer, "a love of the earth and a love of technology could coexist."[41] The high education level of many counterculture youth allowed for a deep engagement with environmental issues. On Denman Island, as Sharon Weaver points out, protesting pollution involved not merely a "not-in-my-backyard" opposition to particular types of economic activity, but also a scientific evaluation of chemical reactions. Emphasizing the body, self-sufficiency, and appropriate technology, counterculturalists fostered new approaches to environmental issues.

EXCHANGES

Stereotypically, the arrival of counterculture youth evoked hostility between them and their neighbours. On Lasqueti Island, BC, local farmers did not appreciate the way that commune leader Ted Sideras allegedly convinced his followers that local livestock was fair game. Sideras was charged with and tried for cattle rustling.[42] On the Sunshine Coast of British Columbia, Wally Peterson, the mayor of Gibsons, complained about the funding that local "longhairs" received from the federal government through the Opportunities for Youth program, suggesting that the money was being used to grow pot rather than potatoes.[43] As Matt Cavers shows, such hostility was fairly common, particularly in the Sunshine Coast region.

In some places in British Columbia, however, counterculture youth encountered people from older generations who had made similar choices in the past. American draft resisters on Malcolm Island met ageing Finnish socialists who knew their Marxist literature much better than the student radicals did. Groups moving into the Kootenays encountered Doukhobors and Quakers who shared similar concerns about the presence of the state, the rejection of war, and the desire to live simply off the land. One neighbour of Doukhobor farmers in southeastern BC recalled, "Their own kids weren't interested in Doukhoborism but here we were, middle-class ex-professionals from California, putting the garden in in the nude, looking for alternatives to materialism and possessive relationships, and working very hard."[44] Likewise, draft resister Marvin Work, who arrived in 1970 in the Kootenays, found ready allies in his Doukhobor landlords, who shared his pacifism.[45] Hippies moving to Hornby Island met the formidable Hilary Brown and her husband Harrison (HB). Hilary had published pacifist and feminist works in Britain in the 1930s before moving to the remote island in 1937. Until her death at ninety-eight, in 2007, she played a key role in founding local co-operative ventures and providing community leadership. The members of La Commune de la Plaine found a perhaps unlikely advocate in their local priest, who preached tolerance and openness to the newcomers.[46]

Thus, despite their beliefs in their revolutionary praxis, counterculture youth often built upon a variety of antecedents, some dating back many decades: socialist perspectives that criticized the inequities of capitalism, pacifist tendencies opposed to militarism, and even long-standing rural distrust of urban centres. Many new commune-dwellers co-operated with and learned from those other groups, and over time they managed to reduce the tensions with other members of the communities.

Perhaps one of the more surprising themes to emerge from this collection is that of the complex links between the counterculture and the state. Many of the individuals displayed tremendous entrepreneurial skills, and in the context of the 1970s this could involve applying for government funds for a range of projects. It is true that funding was relatively accessible at this time—more so than would be the case by the late 1970s, as the financial retrenchment that typified the rest of the end of the century took hold. Prime Minister Pierre Trudeau's close ally Gérard Pelletier served as the minister of state in the early 1970s. Pelletier's department, concerned about the youth-led ferment of the period, offered small-scale funding in Opportunities for Youth and Local Initiatives programs to provide more meaningful work opportunities, he claimed, than a sterile summer job in a government ministry.[47] As Matt Cavers shows, these programs could be fairly lax in standards of application and reporting, and they attracted a lot of local criticism. Indeed, while one arm of the government could dole out grants, other branches, including immigration and police officials, kept tabs on various groups. In 1977, N. S. Fontanne, director of the Intelligence Research and Analysis Division of Canada Immigration corresponded with the Nashville Metropolitan Police Department to acquire information on the famous Tennessee commune "The Farm," because some of its former inhabitants proposed setting up a similar experiment in Lanark, ON.[48]

Government programs involved, to be sure, very small-scale funds, but given the desire to live fairly simply and in areas of the country with inexpensive land values, these funds could make the difference between success and failure. After all, the back-to-the-landers faced the same difficulties almost all utopians confront: how to reconcile spiritual or ideological enthusiasms in a context of collective ownership with a need for the necessities of life. The Ark project on PEI relied on fairly substantial financial support from the federal and provincial governments, having managed to combine concerns about regional development with fears of energy insufficiency, but other projects were built on much smaller sums. The people behind the recycling efforts in Toronto managed to stack application upon application to maximize the subsidy they received, and thus they remained afloat longer than less astute groups. Likewise, cycling activists in Montreal, prospective midwives in the Kootenays, and anti-pollution activists on the West Coast all used small summer funds to bolster their activities.

A further technique that many counterculture activists used effectively was theatre. In other words, they attracted attention for their causes by playing to the media. Oppositional groups have long attempted to achieve public exposure by such methods, and in this way their practices were not much different. Bringing a coffin to the BC Legislature in 1979 to draw attention to pollution on Denman Island or staging a funeral for the putrid Don River in Toronto were not in themselves particularly innovative actions, but they did attract media attention, and they were likely more successful than similar approaches would be in today's oversaturated media cycles.[49] Street theatre could create focal points and moments in which to convey environmental messages, and the theatre of La Commune de la Plaine drew upon *situationniste* models, just as Greenpeace found inspiration in yippie guerilla theatre and the cycling activists in Montreal drew on a range of European and American influences.

Some of the most effective practices involved collaborations with other locals who shared the same appreciation of landscape aesthetics. The most successful attempts to control pollution involved counterculture activists teaming up with local loggers and farmers. In all rural locations, if the young back-to-the-landers had children, they offered the opportunity to keep small schools alive. As Alan MacEachern shows, the counterculture children provided a bridge between the

COLIN M. COATES

newly arrived and the long-standing inhabitants. Increasing familiarity, and labour and other economic exchanges, eventually broke down many barriers. Of course, many back-to-the-landers experienced only a short stay in the countryside, soon returning to the city. Some, like the people involved in Therafields, never really left the city. The farm may have been central to their therapy, but they lived in downtown Toronto. In other locations, the back-to-the-landers raised their children alongside locals, and public schools provided a ground where all groups met—and often worked out their differences. Influences spread both ways, as back-to-the-land children desired bologna while their classmates enjoyed the freedoms the hippie children experienced on their own property. Despite the desire for isolation, the counterculture period also necessarily involved cultural exchange.

CHALLENGES AND LEGACIES

While they may have seen themselves as revolutionaries, in some ways counterculture groups did not challenge the social and racial status quo. Kathleen Rodgers's study of the Vietnam War–resister community in the Kootenays underlines its primarily white and largely middle-class nature.[50] As a number of the chapters discuss, back-to-the-landers encountered neighbours who had never left the land, whether these were farmers in Prince Edward Island or First Nations in the Yukon. David Neufeld explores the complexity of the relations between counterculture youth and Indigenous peoples near Dawson City. In the Yukon, both groups recognized their own countercultural challenges to prevailing opinion and were able to find common ground on some issues, while in many places in the south, counterculture youth embraced ersatz images of Indigenous peoples. One Quebec commune produced its own "native" handicrafts.[51] A meeting of intentional community representatives on Cortez Island, BC, in 1979 began with "Sunrise fires—Indian tobacco ceremony—Sauna and sweats."[52] Indigenous imagery often inspired and informed countercultural worldviews. As Philip Deloria comments in the case of the United States, communalists "promoted community, and at least some of them thought it might be found in an Indianness imagined around notions of social harmony."[53] Many groups were unlikely to connect with First Nations communities close at hand. Feenie Ziner noted the irony of her son's and his friend Buddhi's belief that they had a right to the island where they were squatting:

> How profoundly American both of them were, how middle-class, taking the extravagant promise of their country at face value, converting "I want" into "I have a right to," just like the most avaricious of our fellow countrymen! Neither of them took the exiled Indian population into account in their debate over the right to the land.[54]

One counter-example is noteworthy: Ochiltree, in BC's interior, resolutely engaged not only with the local Aboriginal population, but even more with the poorest Aboriginal street people, creating a joint garden that proved very effective.[55] But partially for this reason, Ochiltree attracted a good deal of local animosity. Locals and the police joined in their dislike of the Marxist commune. Rejecting the idea of private property, Ochiltree members squatted on public lands, and the police attempted to evict them in the 1980s.[56] But Ochiltree was perhaps exceptional among communes in its level of direct engagement and its open defiance of authority.

Communes often remained as strongly gendered as the rest of North American society. Journalist Myrna Kostash points out how communal living experiments failed to challenge gender roles. At La Commune de la Plaine, women went on strike in 1973, withdrawing from the property for a month and leaving the men to care for the children and the household.[57] Commune member Corbeil believes that the male members learned their lesson.

Despite the individualistic, sometimes anarchistic, natures of the communes, they also achieved a degree of institutional fixity. In British Columbia there was even an association of such groups, the Coalition of Intentional Cooperative Communities (CICC). These groups met on a regular basis, every three months, on the site of one of the communes. According to Jim Bowman, the coalition came into existence in response to the then New Democratic Party government of British Columbia. The government was attempting to address issues of communitarian land ownership, but it called an early election in 1975 that it lost, thus ending the chance of passing legislation to allow communes to acquire cheap access to Crown lands.[58]

The CICC newsletters gave space for different communes to discuss their philosophy. Linnea farm on Cortez Island was one of the most ecologically focused communes in British Columbia during this period:

> It is a pilot project focused on developing a harmonious relationship between man and nature in the areas of forest, watershed and eco-farm management. . . . The community members will live close to the land through voluntary simplicity, appropriate technologies, alternate energy and energy conservation. On-going activities are chickens, bees, raw milk dairy, vegetable and fruit production.[59]

For many BC communes, moving back to the land reflected a desire to achieve a simpler existence, although small-scale farming is by no means a straightforward endeavour. As in the United States, the wish for self-sufficiency built on the concerns of many about the military involvements of the American government, fear of environmental degradation, and a general concern that inflation and rapidly rising oil prices would lead to the full-scale collapse of the capitalist system.[60] Communes experimented with alternative forms of energy, sometimes because of a desire to live completely "off the grid" and sometimes only because their choice of an isolated region necessitated it. They also confronted problems of waste disposal, building composting toilets, recycling centres, and "free stores." Hornby Island boasts a particularly famous example, which combines all three in one location, the community having been forced to take action once the local dump was condemned in the 1970s.[61]

Stuart Henderson argues that for some hippies, moving back to the land allowed them to pursue contemporary counterculture lifestyles more fully than did living in Toronto's famous Yorkville neighbourhood, one of the epicentres of the youth rebellion.[62] In general, despite an initial attraction to settling the countryside as a way of escaping mainstream realities, commune-dwellers came face-to-face with the same issues of ecological stewardship that their rural forebears had done. While wishing to establish self-sufficiency, communes also experienced the vagaries of economic life. For instance, in the 1980s the rapid rise in interest rates contributed to the financial difficulties, and ultimately the demise, of La Commune de la Plaine.[63] But the financial failures of some communes should not detract from the long-term impact of their ecological vision.

CONCLUSION

Much of the environmental consciousness that was proposed as counterculture alternatives no longer occupies such a fringe status. The counterculture by no means invented bicycling and recycling, to take two of the issues covered in this collection, but they did popularize both, and they invested strong ecological ethics in the practices. Many current issues can be traced back to their efforts: countercultural support helped to popularize organic farming, controls on harmful chemicals, new attitudes to the human body (particularly in relation to childbirth), concerns about pollution and environmental sustainability, and critiques of technology. All of these have become much more mainstream today than they were in the 1960s. While the counterculture may not have exclusive claim to the parameters of current environmentalist debate, their perspectives created new ethical positions concerning these issues.

The Canadian counterculture was rooted in worldwide youth culture and oppositional stances. While the counterculture emphasized individualities, a larger picture of shared environmentalism developed. Participants engaged with the state—meaning local, provincial, and federal levels in the Canadian context—in an attempt to achieve their aims. Some embraced new technologies, while others eschewed them. They revitalized concepts of land stewardship that remain fixed in agrarian practices.

Like many social movements, the counterculture looked both backward and forward, and its views of the environment reflected both tendencies. Moving back to the land implied returning to a voluntary simplicity, like that proposed by Thoreau in the nineteenth century. John Douglas's rural idyll in northern Ontario in 1967 looked back to a time before electrical water pumps and forward to a spiritual and economic self-sufficiency that entailed a new ecological appreciation. Other members of the counterculture tried to fashion appropriate technologies that would permit sustainable living. As the counterculture foresaw, finding a balance between technology and environment remains one of the most pressing issues facing the world today.

NOTES

1   "Highland Farmstead Is Hippie Heaven for Weekends far from Metro Scene," Toronto Star, 22 September 1967, 2. Thanks to Daniel Ross for this and other references. I would also like to thank Alan MacEachern, director of the Network in Canadian History & Environment, for his comments on this introduction and his support for this project.

2   Mark Vonnegut, The Eden Express (Toronto: Bantam, 1975), 9.

3   Kathleen Rodgers investigates the role of American immigrants in the Canadian counterculture in Welcome to Resisterville: American Dissidents in British Columbia (Vancouver: UBC Press, 2014).

4   Peter Braunstein and Michael William Doyle, "Introduction: Historicizing the American Counterculture of the 1960s and 1970s," in Imagine Nation: The American Counterculture of the 1960s and '70s, ed. Braunstein and Doyle (New York: Routledge, 2002), 10.

5   Robert Hunter, The Greenpeace to Amchitka: An Environmental Odyssey (Vancouver: Arsenal Pulp, 2004), 16

6   Frank Zelko, Make It a Green Peace! The Rise of Countercultural Environmentalism (New York: Oxford University Press, 2013).

7   Judith I. McKenzie, Environmental Politics in Canada: Managing the Commons into the Twenty-First Century (Toronto: Oxford University Press, 2002), 57.

8   Stuart Henderson, Making the Scene: Yorkville and Hip Toronto in the 1960s (Toronto: University of Toronto Press, 2011), 5.

9   Doug Owram, *Born at the Right Time: A History of the Baby Boom Generation* (Toronto: University of Toronto Press, 1996); Bryan D. Palmer, *Canada's 1960s: The Ironies of Identity in a Rebellious Era* (Toronto: University of Toronto Press, 2009); Lara Campbell and Dominique Clément, "Introduction: Time, Age, Myth: Towards a History of the Sixties," in *Debating Dissent: Canada and the Sixties*, ed. Lara Campbell, Dominique Clément, and Gregory S. Kealey (Toronto: University of Toronto Press, 2012), 14–15.

10 Marcel Martel, *Canada the Good: A Short History of Vice since 1500* (Waterloo, ON: Wilfrid Laurier University Press, 2014), 131–36.

11 Feenie Ziner, *Within This Wilderness: The True Story of a Mother's Journey to Rediscover Her Son* (New York: W. W. Norton, 1978), 7.

12 Vonnegut, *The Eden Express*, 33.

13 "Back to the Land," *Take 30*, CBC Television, broadcast 25 April 1969, accessed 1 July 2014, http://www.cbc.ca/archives/entry/back-to-the-land.

14 Terry Allan Simmons, "But We Must Cultivate Our Garden: Twentieth-Century Pioneering in Rural British Columbia," (PhD diss., University of Minnesota, 1979), 93–94.

15 Section on "Communes" from the "Revolutionary Hippy Manifesto" (1978), file 3, box 1, Intentional Community Collection, RBSC-ARC-1273, UBC Library Rare Books and Special Collections, Vancouver (hereafter Intentional Community Collection).

16 Quoted in Katherine Gordon, *The Slocan: Portrait of a Valley* (Winlaw, BC: Sono Nis, 2004), 238.

17 Michael Egan, "Shamans of the Spring: Environmentalism and the New Jeremiad," in *New World Coming: The Sixties and the Shaping of Global Consciousness*, ed. Karen Dubinsky et al. (Toronto: Between the Lines, 2009), 296–303.

18 In Marc Corbeil's words, "l'autarcie la plus totale possible." "C'était une démarche en recherche d'une vie saine, en contact avec la nature, avec les autres et avec soi-même, où le travail serait régénérateur et où les plaisirs du corps et de l'esprit auraient leurs droits." Marc Corbeil, *L'Utopie en acte: La Commune de la Plaine* (Rimouski: Université du Québec à Rimouski, 1990), 18, 45, 55. (All translations by the author.)

19 Grant Goodbrand, *Therafields: The Rise and Fall of Lea Hindley-Smith's Psychoanalytic Commune* (Toronto: ECW, 2010), 1.

20 Lea Hindley-Smith, Stan Kutz, Philip McKenna, and bp nichol, "Therafields" *Canadian Forum* 52 (January 1973), 17.

21 Zena Cherry, "Colborne Honored," *Globe and Mail*, 22 August 1978.

22   For instance, Peter Coyote, excerpt from "Sleeping Where I Fall," in *The Counterculture Reader*, ed. E. A. Swingrover (New York: Longman, 2004), 47; Christopher Gair, *The American Counterculture* (Edinburgh: Edinburgh University Press, 2007), 220; Gretchen Lemke-Santangelo, *Daughters of Aquarius: Women of the Sixties Counterculture* (Lawrence: University Press of Kansas, 2009), 182.

23   Quoted in Mark Kurlansky, *1968: The Year that Rocked the World* (New York: Random House, 2004), 120.

24   Adam Rome, "'Give Earth a Chance': The Environmental Movement and the Sixties," *Journal of American History* 90, no. 2 (2003): 525–54.

25   Keith M. Woodhouse, "The Politics of Ecology: Environmentalism and Liberalism in the 1960s," *Journal for the Study of Radicalism* 2, no. 2 (2008): 77.

26   F. Kenneth Hare, "Our Total Environment," in *Crisis: Readings in Environmental Issues and Strategies*, ed. Robert M. Irving and George B. Priddle (Toronto: Macmillan, 1971), 346.

27   McKenzie, *Environmental Politics in Canada*, 62.

28   Zelko, *Make It a Green Peace*, 10–32, 77.

29   Novia Carter, *Something of Promise: The Canadian Communes* (Ottawa: Canadian Council on Social Development, 1974), 13.

30   Mark Frutkin, *Erratic North: A Vietnam Draft Resister's Life in the Canadian Bush* (Toronto: Dundurn, 2008), 71.

31   "Les Communes," *Tout le monde en parlait*, Radio-Canada, broadcast 8 June 2010, accessed 1 July 2014, http://ici.radio-canada.ca/emissions/tout_le_monde_en_parlait/2010/Reportage.asp?idDoc=112586.

32   Gregory Baum, interview by Matt McGeachy, 25 January 2008, transcript. My thanks to Matt for sharing this transcript.

33   "Les Communes."

34   Quoted in Sharon Weaver, "First Encounters: 1970s Back-to-the-Land Cape Breton, NS and Denman, Hornby and Lasqueti Islands, BC," *Oral History Forum d'histoire orale* 30 (2010): 19.

35   Bill Osgerby, *Playboys in Paradise: Masculinity, Youth and Leisure-Style in Modern America* (Oxford: Berg, 2001), 167–72.

36   "Les Communes."

37   Frutkin, *Erratic North*, 210–11.

38   E. B. Demerchant, "US Youth Colony Rediscovers Life on Old NB Farm," *Telegraph-Journal* (St. John), 6 May 1970. Thanks to Greg Marquis for this reference.

39   "Les communards avaient l'impression erronée que les modes de production antérieurs étaient moins compliqués" (Corbeil, *L'Utopie en acte*, 22).

40 On the role of counterculture environmentalists in popularizing "appropriate technology," see Andrew G. Kirk, *Counterculture Green: The* Whole Earth Catalog *and American Environmentalism* (Lawrence: University Press of Kansas, 2007).

41 Walter Isaacson, *The Innovators: How a Group of Hackers, Geniuses, and Geeks Created the Digital Revolution* (New York: Simon & Schuster, 2014), 272.

42 Larry Still, "Island Cattle 'for God's Children," *Vancouver Sun*, 8 June 1972, 1–2.

43 "Sunshine Coast Communes Arouse Local Ire," *Vancouver Sun*, 28 June 1971, 33.

44 Quoted in Myrna Kostash, *Long Way from Home: The Story of the Sixties Generation in Canada* (Toronto: Lorimer, 1980), 119.

45 Jessica Squires, *Building Sanctuary: The Movement to Support Vietnam War Resisters in Canada, 1965–1973* (Vancouver: UBC Press, 2013), 73.

46 Corbeil, *L'Utopie en acte*, 15.

47 Gérard Pelletier, *L'Aventure du pouvoir, 1968–1975* (Montreal: Stanké, 1992), 142–43.

48 N.S. Fontanne to Lt. Thomas E. Cathey, Field Operations Bureau, Nashville, 10 February 1977, vol. 1405, R1206-238-2-E, Intelligence Division, Library and Archives Canada, Ottawa.

49 Ryan O'Connor attributes the success of Pollution Probe— which staged the Don River's mock funeral—to its centrist, and therefore non-countercultural, politics (*The First Green Wave: Pollution Probe and the Origins of Environmental Activism in Ontario* [Vancouver: UBC Press, 2014], 52–56, 170–73). The similarity of some of this group's strategies to those used by more radical groups is striking.

50 Rodgers, *Welcome to Resisterville.*

51 "Les Communes."

52 Coalition of Intentional Co-operative Communities, "Summer-fullmoon-gathering," *Open Circle* (newsletter), 1979, file 2, box 1, Intentional Community Collection.

53 Philip Deloria, "Counterculture Indians and the New Age," in Braunstein and Doyle, *Imagine Nation*, 160.

54 Ziner, *Within This Wilderness*, 106.

55 Justine Brown, *All Possible Worlds: Utopian Experiments in British Columbia* (Vancouver: New Star, 1995), 68–69.

56 Don Hunter, "Quit Cariboo, 'Hippies' Told," *The Province*, 18 February 1985.

57 Corbeil, *L'Utopie en acte*, 16.

58 "Coalition of Intentional Cooperative Communities" *Communities: Journal of Cooperative Living* 47, file 5, box 1, Intentional Community Collection .

59 "Turtle Island: A Land Stewardship Society," *Open Circle* (newsletter), 1979, file 2, box 1,

Intentional Community Collection.

60 Jinny A. Turman-Deal, "'We Were an Oddity': A Look at the Back-to-the-Land Movement in Appalachia," *West Virginia History*, new series, 4, no. 1 (2010): 1–6.

61 *Everything Old Can Be New Again*, directed by Dale Devost (Hornby Island, BC: Outer Island Productions, 2006), DVD.

62 Henderson, *Making the Scene*, 269.

63 Corbeil, *L'Utopie en acte*, 31.

SECTION 1:

ENVIRONMENTAL ACTIVISM

Back-to-the-Land Environmentalism and Small Island Ecology: Denman Island, BC, 1974–1979

Sharon Weaver

On August 14, 1979, Leslie Dunsmore of Denman Island testified before the Herbicide Appeal Board, arguing against the use of Tordon 22K by Weldwood of Canada. In the *Denman Rag and Bone*, a local newsletter, editor Des Kennedy reported on Dunsmore's impressive performance:

> I can't see anyone in the room who isn't listening intently. Her presentation moves like a just-honed scythe through dry grass. She discusses her own livelihood as a beekeeper, the potential for contamination of domestic water supplies, the soil classifications and topography of the area, the properties and hazards of Picloram, forest management alternatives and the limitations of the licensing and appeal processes. Her text is laced with references to experts, commissions of inquiry and scientific studies.[1]

Following her brilliant testimony, Weldwood's cross-examination faltered and sputtered out, reported Kennedy. Dunsmore, like Kennedy, was a back-to-the-lander who had settled on Denman Island within the previous five years. Both came from large urban centres where they had obtained university degrees, and while their degrees were not in science, their education gave both the confidence to question authority and to do their own research. Kennedy reported that "being at the Hearing made one feel proud and happy to be from Denman, to have neighbours of such skill and dedication." This fight against the spraying of herbicides was one of a series of environmental struggles in which Denman Islanders had engaged over the previous six years. Through local media and debates, back-to-the-landers on Denman Island confronted very local environmental pressures, and in a number of cases—and despite the odds against them—they succeeded in changing decisions and regulations. Their successes were frequently predicated on their ability to engage the concerns and energy of other islanders.

SMALL ISLAND ECOLOGY

Small island ecological systems have been at the cutting edge of environmental concerns and science since at least the seventeenth century. Resource depletion on small islands becomes evident long before it can be detected on the mainland and thus serves as a warning, much like the proverbial canary in the coal mine, to unsustainable draws on natural resources. Their small, bounded geography allows no easy solution to the unexpected collapse of a resource. Historian Richard Grove noted that island contexts led to very early efforts to mitigate environmental change. Both French scientists on Mauritius and English scientists on St. Helena alerted their metropolitan governments in the eighteenth century to the threats posed to the islands' viability by the unrestricted use of resources such as timber, fruit, and water.[2] Even in less isolated locations, ecological impacts are often much more visible on islands than on continents.

Ironically, Denman and other small Gulf Islands located on one of the world's wettest coasts face serious water problems.[3] Sitting in the rain shadow of Vancouver Island's mountains, they are arid, with just half the rainfall of the Vancouver region.[4] In years of light winter rainfall, groundwater is not replenished, and summer shortages are more likely. Because groundwater from wells on Denman is the principal source of domestic and agricultural water, in addition to the two small lakes, the summer rise in population due to cottagers and tourists exacerbates water problems. Overuse of aquifers can lead to saltwater intrusion;[5] by the 1970s, this potential threat to the water table had become a source of concern for most islanders, old and new settlers alike. As well, the cutting and hauling of timber contributed to water degradation, and with increased settlement, the impact of logging on water resources grew more alarming.

In the 1970s, environmental unease among North Americans intensified, moving from the margins to the mainstream. The emergence of the environmental movement provided ordinary people with the sense that they could have a say. Ecological disquiet often motivated back-to-the-landers, with many arguing that their way of life testified to their environmental concerns: gardening without pesticides, herbicides or artificial chemicals; heating with wood; building with local materials; and opting out of consumer culture all demonstrated their environmental credentials.[6] By moving to relatively remote areas such as the Gulf Islands and Cape Breton, back-to-the-landers were trying to escape the long reach of capitalist, industrial society. However, they quickly discovered that they could not entirely break free from it. Even those for whom the environment was not a primary motivation were quick to defend a right to clean water and clean air.

Known for their extraordinary beauty, unique ecosystems, and biological diversity, the Gulf Islands had come under increased developmental pressure during the 1960s.[7] Growing public alarm over uncontrolled development, possibly beyond their carrying capacity, led W. A. C. Bennett's Social Credit government to impose restrictions in 1969, limiting island subdivisions to lots no smaller than ten acres. Previously, a lack of planning for the islands had arisen out of the fact that British Columbia provided only a reduced framework for local governance outside of municipalities. With the creation in 1965 of twenty-nine regional boards spanning the entire province, citizens living in rural districts obtained a limited form of governance—which was clearly inadequate, as district boundaries combined municipalities with surrounding unincorporated areas.[8] Because population determined the voting weight of each elected member to a regional board, this usually resulted in the islands having little to no individual representation on these boards. As an example of scale, in 1981, the Regional District of Comox-Strathcona had a total population of 68,621; within it, Denman Island's population was 589, and Hornby Island's, 686.[9] With next to no input from island residents on any of the boards, little time or effort was devoted to island issues.[10] As a result, many islanders viewed the imposition of policies by the larger region as "illegitimate uses of political power," and the problem "resulted in considerable dislike for regional district government in some rural areas."[11] Acknowledging "the special planning needs of island environments," the New Democratic Party government held meetings in 1973 on the thirteen most populated islands, seeking input on how best to create a governing structure for those islands. As a result of these consultations, the Islands Trust Act was proposed and enacted in 1974.[12] The Islands Trust staff act as a regional board for the thirteen islands that fell under the new legislation, with two elected trustees from each island, who, as of 1979, then elected a chair and vice-chair.[13] The freeze from further subdivision into parcels smaller than ten acres continued until a community plan could be developed on each island. It was hoped that the new legislation would put in place controls to preserve and protect the rural qualities of the islands, "given the uniqueness of island environments, the insignificance of island concerns in regional districts and the sense of community that exists among island residents."[14]

Like other islands in the Strait of Georgia, Denman experienced a rapid increase in population beginning in the late 1960s and continuing throughout the 1970s.[15] Many of these newcomers were young, often well educated, and in search of a retreat from uncontrolled growth, industrialization, and pollution. A large proportion came from the United States, where debates about the environment were gaining public attention. The population on the island at this time consisted of, in addition to the newcomers, descendants of the original European families who had settled on the island in the latter part of the nineteenth century along with recent retirees, many of whom had summered on the island and then chosen to make it their year-round home.

THE *DENMAN RAG AND BONE*

On Denman, the vulnerabilities of small island ecologies soon brought the back-to-the-landers into the open, as the best hope for mitigation depended on both disseminating information and generating activism. In 1974, Des Kennedy and Manfred Rupp began the *Denman Rag and Bone*, a newsletter of local environmental concerns. Kennedy later stated in an interview that

> myself and a number of others, sort of more politically oriented people, very quickly realized . . . the islands were totally ripe for plucking by land speculators and developers and stuck our noses in and said that's not what we want to have happen here . . . and that's where the *Denman Rag and Bone* sprang out of, that desire to mobilize the community around the need for, at least from my perspective, for that kind of vigilance, because you could see it start to happen, whether gyppo loggers coming in and just butchering the place, . . . [or] land speculation and development.[16]

Conceived of and launched as a community newspaper, the *Denman Rag and Bone* encouraged islanders to communicate with one another. In the span of five and a half years, it reported on numerous issues that constituted a threat to the island's ecosystem and that local people tackled. Concerns included inappropriate recreational use of Chickadee Lake, road maintenance, and the impact of summer tourism on island capacity, all of which required wider discussion. The threat of contamination posed to the water table by excessive subdivision development and the proposed herbicide spraying by both BC Hydro and Weldwood were particularly alarming to Denman Islanders. With each of these environmental concerns we see how the repeated stresses on Denman Island's small ecosystem became inescapable and how residents were forced to address them and, in the process, reconfigure social alignments.

Fundamentally a back-to-the-land source, the newsletter—which was published from May 1974 until August 1979—included the voices of others as responders or guest contributors. It featured local, regional, and provincial developments that might affect the island, thereby fostering a greater sense of community. Small as Denman Island was (the size of Manhattan Island, with 379 permanent residents in 1976[17]), gossip and informal networks were inadequate for the dissemination of complex information, especially that needed for informed voting. My research, both in reading the newsletter and through conducting personal interviews, made it clear that islanders felt that bylaw decisions and development policy were controlled by a few key individuals, who were unaccustomed to sharing information with fellow residents. A newsletter delivered to each mailbox about six times a year seemed the best way to share information, create informed discussion, and perhaps circumvent the established powerbrokers. The *Denman Rag and Bone* was delivered free of charge up to and including issue number 25, after which the cost was twenty-five cents per issue in the general store, or ten issues delivered on the island for four dollars (five dollars off the island).[18]

While environmental matters were an important part of the content, a typical issue included artwork, poetry, short fiction, recipes, gardening advice, children's or school pages, editorials, letters to the editor, and occasionally pieces originating on other islands or elsewhere in the province. A page or two under the heading of either "Rumours Galore" or "Bits & Pieces"[19] included short paragraphs about individuals on the island, upcoming meetings, ongoing issues, and almost always a paragraph on local road conditions. Notices appeared about the food co-op and forthcoming meetings of the Ratepayers' Association, the Fire and First Aid Committee, and the Recreation Committee. The newsletter also contributed to island history, frequently in the form of an interview with a long-time resident. Women contributed to the newsletter as both writers and workers.

SHARON WEAVER

Their contributions covered topics that were typically associated with women's concerns, such as children, food, and gardening, but they, as well as a few brave men, also wrote about women's changing role in society, motivated by the growing awareness created by the feminist movement. Fifteen to twenty people contributed content, while about ten volunteers typed and laid out the text and ran the Gestetner to produce each issue. Then, to ensure a wide readership, volunteers delivered the newsletter to (in the beginning) every island mailbox.

For the first two years, the strongest voices in the newsletter belonged to its two founders, Des Kennedy and Manfred Rupp. Kennedy was born in Liverpool, England, in 1945 and moved to Toronto with his family at age ten. He "then spent eight years in a series of monastic seminaries in the Eastern United States, studying for the priesthood," which he left in 1968 to move to Vancouver; there he met his wife, Sandy, while they were both employed as social workers.[20] As a former monk, Kennedy was drawn to the quiet and seclusion of the woods, as was Sandy, and so together they spent their weekends, "weather permitting, out in the woods somewhere . . . camping." Rather than have to "drag" themselves "back to the city" every Sunday evening, the couple bought land on Denman Island in 1971 and took up permanent residency in 1972. According to Kennedy, "I had an ambition to be a writer and so we were looking for simplicity, frugality and quiet."[21] He planned to support himself as a professional writer and, with his exceptional gardening skills, wrought a wondrous transformation of his eleven acres on Denman Island and then wrote popular books on gardening, among other genres.

Manfred Rupp and his wife, Marjo Van Tooren, bought land on Denman Island in 1969 and were some of the earliest to arrive of the back-to-the-lander group. Born in Germany in 1931, Rupp recalled one of the more formative experiences of his early years:

> Growing up in Germany as a teenager I spent my holidays hitchhiking to what was then [a] very popular international work camp where some organizations, in my case it was a branch of the Quakers, . . . set up camps in places where

there was need. . . . In one case we went into Holland after a flood and did cleanup, same in Austria, or in Norway we blasted a road, . . . this kind of stuff. That was really high times . . . that made me begin to see how nice it is when people get together and manage to organize for a common purpose.[22]

Rupp became a teacher, and to his surprise his immigration application to Canada was accepted. Arriving by boat in 1958, he picked fruit in southern Ontario to repay his fare and then found administrative work at the University of Alberta in Edmonton. He saved enough to buy a tea house in Horseshoe Bay, BC, hoping also to make pottery, but, in his words, the business "flopped" because "Horseshoe Bay is hamburger country." After owning an art gallery in Vancouver for a few years, the couple had recouped enough money to buy a small property on Denman Island, where Rupp thought he might finally attempt to live in a way that reflected the inspiration he had felt working cooperatively with the Quakers as a teenager in Germany:

> We didn't necessarily, primarily, go back to the land; for us we were four couples living in Vancouver . . . looking for land. . . . It was a real attempt, failed attempt I might add, to invent a co-operative lifestyle.[23]

When his son was about to enter school, Rupp decided to move his family to Germany, where his son would learn German and become acquainted with his relatives. With Rupp's departure, Kennedy became chief editorialist and frequent "Bits & Pieces" columnist. Although his voice and politics tended to dominate the *Denman Rag and Bone*, the viewpoints of others also appeared regularly—and they were not always in agreement with Kennedy. The newsletter conveyed countercultural approaches to island living, but it also attempted to address the concerns of the entire population, which were frequently discussed at local community meetings.

Ratepayers associations and community clubs were important forums for debate on all of the islands. Because of the small populations of the areas involved, many of the meetings of these groups resembled old-fashioned town meetings where issues were discussed by a large proportion of the community, in contrast to the professionalized, superficial, mass media presentations so often found in the larger urban municipalities.[24]

The *Denman Rag and Bone* regularly recorded concerns and debates raised in the ratepayers meetings on Denman. Not surprisingly, the back-to-the-landers and some of the long-time residents often represented different perspectives. With regard to the Islands Trust legislation, the back-to-the-landers, alarmed over resource depletion and uncontrolled development, welcomed the possibility of greater control that the new legislation represented. Back-to-the-land settler values often clashed with those of the other stakeholders on the island. Forming the bulk of the back-to-the-land settlers, the baby boom generation, unlike preceding generations, had the luxury of a relatively peaceful existence coupled with financial security, which allowed them to focus on issues of equality and environmental protection.[25] The larger landowners, in contrast, worried over what the new act might mean for their ability to manage their property, including their right to subdivide should they wish. Inhabiting the middle ground in the debates were the recent retirees and perhaps a substantial number of island residents.

How representative of the back-to-the-land opinion was the *Denman Rag and Bone*? Given the large number of contributors to both its production and its content over the years of its existence, the substance of the letters in response to its editorials, and the thirty interviews I conducted between 2005 and 2008, it seems safe to conclude that the newsletter reflected back-to-the-land opinion accurately. As for land development issues, certainly some back-to-the-landers held land cooperatively, which meant their interests in subdividing that land might have differed somewhat, but not substantially. The key issues with development, according to the *Denman Rag and Bone* were scale and resource use, including, in particular, depletion of the water table and potential bottlenecks at the ferry terminal. As Kennedy phrased it in the first issue, "logging and road widening . . . along with rip-off subdivisions do far more to destroy the 'unique amenities and environment' than does some poor citizen erecting a supplementary outhouse."[26]

When the *Denman Rag and Bone* was founded, back-to-the-landers had been on the island for no more than five years. With the imminent passage of the Islands Trust Act in June 1974, the newsletter provided timely information to islanders about the many issues surrounding this piece of legislation.[27] At this point the actual control each island would have in developing its own bylaws and policies was yet to be determined, as were the duties of the elected trustees. An editorial in the *Denman Rag and Bone* read, "we believe that this island is at a critical point in its current stage of development. How we approach that point will in large measure determine the kind of place it will become."[28] Writers for the *Denman Rag and Bone* felt that every citizen living on the island should have a clear understanding and full awareness of the many issues facing their surroundings.

Despite initial enthusiasm for the new legislation and the possibility of thereby gaining increased control over the pace of development on the island, the *Denman Rag and Bone* did not offer an unmitigated endorsement:

> Lest our apparent editorial bias in favour of the Trust Act be misconstrued to mean uncritical acceptance, we repeat certain questions asked in our first issue. An obvious one: while the act clearly intends to muscle into Regional Board territory . . . it seems to avoid very adroitly stepping on the toes of fellow ministers such as Highways and Forestry. What influence will the Island Committee have on the policies of those departments, as they affect the islands?[29]

Whether these issues concerned the two lakes on the island, road widening, herbicides, tourism, the hydro company, or development, any one of them had the ability to negatively impact the environment of Denman Island and the quality of life there. Fundamentally, water quantity or quality underlay all of the issues.

By the end of 1979, when the *Denman Rag and Bone* ceased publication, much had been accomplished in averting some of the more flagrant disregard of bylaws by developers. Logging companies and even government departments had learned to consult islanders before unilaterally initiating action on the island. Ratepayers meetings were much better attended, a new environmental group had been created—Alternatives for Community and Environment (ACE)—and a bylaw support group had been formed to research and support the island's zoning bylaws.

CONTESTED VISIONS

CHICKADEE LAKE

One of two lakes on the island, Chickadee Lake was a source of contention between new and longer-term residents, and it was discussed frequently in the *Denman Rag and Bone* during its early years of publication. Tension built over the fact that many of the newer residents (by no means all) enjoyed nude swimming in the lake, which, not surprisingly, offended some of the original islanders. The most vocal of those who took offence was Wes Piercy, president of the Recreation Committee at the time. Piercy, like many local islanders, had fond memories of swimming at Chickadee Lake as a child and wanted his grandchildren to be able to enjoy the lake as he had, and nude swimming by a bunch of "hippies" did not fit with his vision. Weldwood of Canada, a subsidiary of United States Plywood, had acquired a portion of the land adjacent to the lake, which it managed as a tree farm, and ostensibly out of civic duty—but more likely for strategic reasons—had proposed a picnic-site development at the lake. This project involved "opening it up." The company proposed clearing away trees and brush, hauling in loads of sand, creating a parking lot, and adding garbage cans and picnic tables. This proposal seemed the perfect solution to Piercy; it would open the site to greater public scrutiny and effectively reduce the likelihood that nude swimmers would find the lake an attractive location.

At a July ratepayers meeting, Weldwood representative W. A. Hopwood informed the community that the company planned to construct a logging road nearby, asserting that "no permit is needed [for the road], we just build it." Furthermore, while a tree farm is normally defined as an area of land managed to ensure a continuous supply of wood for commercial production, Hopwood informed those present that a "tree farm relates to tax status, not forestry status."[30] Close proximity of a logging road would hardly enhance enjoyment of the new picnic site, to say nothing of the obvious environmental impacts to the lake and its water quality. Rupp, in a *Denman Rag and Bone* editorial, questioned whether the increasing demand for recreational access to the lake was incompatible with the necessity to preserve a potable freshwater supply.[31] Beyond the back-to-the-land crowd, according to a later editorial, the Denman Island Planning Study of 1971 had recommended the lake be preserved in its "natural" state, as did the regional district's *Evaluation of Proposed Greenbelt Sites*, which went a step further and suggested acquisition of the lake and its surrounding land to "prevent developments harmful to wildlife." Finally, the proposed community plan recommended that the lake be "preserved in its natural state and not opened up for tourist use."[32]

Peter McGuigan, Harold Walton (president of the Ratepayers' Association[33]), and others visited the lake to provide recommendations for Weldwood. They discovered that the company had already begun to dump sand in a "tasteless nature" but had been forced to desist by adjacent property owners. When McGuigan reported their findings at the next ratepayers meeting, "instead of discussion," he encountered "loud shouts of comic book treason . . . from the back bench," ending "with a suggestion that those who didn't agree [with the company's actions] should leave the island." In fact, McGuigan reported, he had wanted to suggest that a *small* picnic site be developed on Chickadee Lake, providing attention was paid to the ecology of the lake and its long-term value to islanders and the people of the province.[34] Finally, the Ratepayers' Association concluded that none of the sites suggested by Weldwood was acceptable to the community.

Thinking the matter settled, those attending a subsequent ratepayers meeting were surprised to discover that Piercy, of the Recreation Committee, had made a behind-the-scenes agreement with Weldwood approving the site that had been rejected by the ratepayers. Asked by Piercy to support his action, Walton insisted on a meeting between Weldwood, ratepayers, and the Recreation Committee.[35] Apparently no resolution was achieved, because a year later, Walton, in his role as one of the newly elected island trustees, asked the Islands Trust to intervene to help save the lake, as Weldwood had made the arbitrary decision to go ahead with building a logging road a mere 110 feet from the lakeshore. This meeting was chaired by the Islands Trust manager, Judy Parr, and although it was reported that Weldwood managed to sidestep the entire matter, the final result was that a small picnic site was established at the lake, while the proposed logging road was averted due to the near consensus against it among islanders. Concern over roads and their impact on the watershed was not limited to logging roads; however, the impact of road widening at the expense of waterways or aesthetics was not as readily evident and required more time and discussion.

ROAD MAINTENANCE

Roads and the decision-making process within the department of highways endlessly frustrated many islanders, who found themselves pondering whose authority determined what seemed a continuous round of widening and grading, and what logic was behind these decisions. Furthermore, the heavy equipment used had to be ferried over for each job, occupying valuable ferry space. Manfred Rupp editorialized on why the roads kept changing and who made the decisions that led to the "dreadful mess we see spreading along our roads." He argued,

> if we didn't know any better we would have to conclude that what we see is one bunch of machines preparing the way for another bunch of machines, with no observable interference from human intelligence. You talk to any of the

higher-ups in the Highways establishment and they will in-variably justify their heavy-handedness by referring to the needs of machines.[36]

Rupp's frustration lay in the fact that decisions about roads in unincor-porated areas were made by the provincial Ministry of Transportation and Highways, while actual road maintenance work involved little interaction between the local population and the decision makers.[37] The Denman Island road foreman, Cliff Grieve, stated that road maintenance and budgetary decisions were made in Courtenay, on Vancouver Island.[38] Taking matters into their own hands, islanders formed a roads committee charged with investigating problems. This committee reported its findings to the Ratepayers' Association: in ad-dition to excessive widening, roadside vegetation had been destroyed, topsoil removed, cliff-top vegetation uprooted, potentially leading to erosion, and the old cedar fences for which Denman was known were often carelessly battered down. Kennedy noted in an editorial that "road 'improvement' is a touchy subject hereabouts, especially with certain local statesmen who tend to go into an irrational froth when the subject is raised." The reference to "local statesmen" was no doubt directed at the two elected trustees, Harold Walton, former president of the Ratepayers' Association, and Marcus Isbister, whose family had lived on the island for generations. From the back-to-the-land perspec-tive, winding, tree-lined roads, seldom found in the city or suburbs, formed a fundamental part of the island's unique attractiveness, and the highways department's uniform approach of standardizing roads across the province was ruinous to the Gulf Islands. In Kennedy's opinion, the Islands Trust had not "demonstrated much leadership" on the issue, and he "hoped that Ratepayers could fill the breach"; in fact, Kennedy suggested that, "if you're interested, Ratepayers is where it's happening."[39] This last quip marked a significant change in the makeup of the association, in fact, as it had only been two years since the Ratepayers' Association had seemed to exclude the back-to-the-landers. Now, Kennedy was proud of the fact that discussion was already "under way concerning the construction of pathways for pedestrians, cyclists and horse riders along main roadways."[40]

On a more humorous note, beaver ponds were found to be another casualty of the highways department, which found the ponds and their inhabitants a nuisance to road maintenance. The government's solution was often to fill in the pond or at least the portion deemed necessary to the road, culvert, or bridge. Not surprisingly, this caused yet more friction between the department and islanders, or at least those islanders with strong notions about watershed preservation and equally strong beliefs that roads and bridges tended to be overbuilt. Following a recent bridge-building project along his road, Kennedy reported that the department of highways had sent

> a stiff reprimand to the residents of Pickle's road who were accused of feeding fertility pills to the beavers in a baldfaced attempt to undermine the Highways Dept bridge-building endeavour. The Catholic members of the accused were particularly shaken because pills (to stop or multiply) are a no-no in Papist circles. A Papal Bull threatening excommunication to anyone counselling fertility manipulation of any kind (other than the Rhythm Method—never very popular with beavers) has been posted on the new bridge which, despite a certain appearance of overkill, has been well constructed and will, we hope, serve Island needs for years to come.[41]

Road debates highlighted both aesthetics and watershed worries and fed into discussions of island tourism.

TOURISM

Although tourism was the third largest industry in BC, it remained an industry islanders loved to hate—and on this point, both newcomers and old-timers could agree. As Kennedy argued, "they're just folks away from home," but "too often they're a pain in the ass."[42] He also quoted a 1971 article by Doras Kirk, who was born and raised on the island, to demonstrate, presumably, that it was not just island new-comers who harboured antipathy to tourists. Drawing from a survey, Kirk had stated that "Denman Island Ratepayers want Denman Island for Denman residents, not tourists."[43] The survey had shown that residents believed a significant influx of tourists would create serious problems, including inadequate supervision of parks, increased fire hazards, and garbage disposal issues. Both of Denman's lakes, licensed for water supply, would be unsuitable for tourists, as the lakes were small and vulnerable to overuse. As well, city folks, accustomed to an endless supply, would likely use excessive amounts of water, affecting Denman's water table. Finally the report also noted that an increase in tourist traffic would stir up more dust from unpaved roads. Five years later not much had changed; a great number of tourists still posed problems due to inadequate facilities, Kennedy argued. Summer homes were "popping up on East road . . . like chickweed amongst the cabbages." They seemed to go hand in hand with "the grating whine of trail-bikes" that was "becoming as familiar as long ferry line-ups." Kennedy's description continued: "Sahara-size dust-storms chase speeding cars down gravel roads. And it leaves a trail of debris behind it. Beer bottles, candy wrappers and other crap begin to litter the shoreline of our beloved Chickadee Lake."[44] Much as transient hippies were initially lumped in with back-to-the-landers due to their appearance, it is possible that summer residents, who returned year after year and often owned property on the island, were being unfairly lumped in with the casual tourists.

Nevertheless, although some summer residents probably made an effort to partake in community events, for the most part there was a sense that their presence on the island was only fleeting. Some stayed for as little as a week or two, thus contributing little to the community other than increases in traffic congestion and property taxes.[45] The community and the trustees, argued Kennedy, had to come up with innovative ways to accommodate tourists and, especially, consider what kind of tourist they wanted to attract. In this case, the Islands Trust was mandated with a dual and somewhat conflicting mandate. On the one hand, it was to preserve the rural flavour of the islands, but on the other, it had to balance that with its mandate to preserve the islands for all residents of the province. Tourism that involved hiking, bicycling, horseback riding, sailing, and kayaking were activities that best fit the bill, suggested Kennedy. A certain degree of consensus seemed to exist among islanders on this matter. The survey reported by Kirk demonstrated that antipathy toward tourists was not exclusive to back-to-the-landers. Kennedy broadened the issue to include cottagers. Islanders were particularly sensitive to development and created strict bylaws to protect island resources. Any flouting of these bylaws by an outsider brought all islanders together in opposition.

OVERUSE AND ABUSE OF ISLAND WATER

The *Denman Rag and Bone* was highly effective at galvanizing support and organizing protests concerning water resources, framing such issues as outsider interference with island bylaws or as serious threats to health.

GRAHAM LAKE SUBDIVISION

Frank Rainsford, an off-island developer, was at the heart of a long-running and contentious issue on the island. His proposed subdivision on Graham Lake, known as Seaview Estates, was to consist of twenty-two lots, but Rainsford later sought approval for fifty-three lots. Though it is unclear whether the original twenty-two lots had already been formally approved, many islanders considered fifty-three lots an overly dense subdivision. "In 1976, Trustees Walton and Isbister received Island support to avert [this] flagrant disregard of Denman's Community Plan, Trust objectives, and local land development restrictions," reported Paul Bailey.[46] Disallowed by the province in 1976, the proposal reappeared in 1979; this time the senior approving officer for the provincial government, Don South, stated that he saw no reason to prevent the development.[47] "Two years ago," according to Bailey, "the same man had told Denman representatives that he would take Rainsford to court rather than offer final approval status." As early as 1974, Harold Walton, then president of the Ratepayers'

Association, had reassured fellow islanders, with regard to the impending Trust legislation, that "Denman Island is one of a few islands fortunate enough to have subdivision and zoning by-laws already in effect." He also noted at the time, "as Municipal Affairs Minister [Jim] Lorimer has stated unequivocally, the new legislation will not be used to either change or circumvent these existing by-laws." Additionally, the minister had in the past vetoed the wishes of islanders, whereas the new legislation offered recourse to the courts. In a letter dated May 30, 1974, Lorimer also stated that "it is the intent that the formation of the Trust will actually give the people of the Islands more say in their own affairs."[48] But a change in government with the election of Bill Bennett's Social Credit Party had apparently negated these gains.

Rainsford's proposal and the province's response left islanders justifiably furious at this clear contempt for their bylaws. "To combat this recent insult to public and Island political and legal sensibilities," Bailey wrote, "Denman Trustees Glen Snook and Harlene Holm have contacted the news media to spread the word."[49] The local television station, CHEK-TV, "featured a short but to-the-point interview with Harlene [Holm] and C.B.C. aired the story twice on their 'Good Morning Show.'"[50] Both trustees later received phone calls and letters pledging moral and even financial support should the issue become a class action suit. At the same time they learned that other islands had had similar problems with the same developer. The only people they did not hear from were Don South or Highways Minister Alex Fraser.[51]

This issue was finally resolved, but only after islanders took their collective protest to Victoria. On July 9, 1979, eighty islanders held a bylaw funeral in front of the Parliament Buildings. The islanders, young and old, marched in two-by-two formation; a protester at the front held an RIP sign representing the death of their bylaws, while the rest followed quietly. Next came a drummer beating a dirge, and those bringing up the rear carried a coffin that contained a copy of the island's bylaws. As they stood solemnly tossing bylaws into a bonfire, "word came . . . that the Minister of Highways had consented to meet with a smaller group in the near future."[52] The public demonstration by so many islanders, coupled with the resulting media attention, had the desired impact: the province rescinded approval for the Seaview Estates development.[53] Well-organized street theatre had persuaded the provincial authorities to respect local attitudes toward development proposals. The issue of respect for island bylaws by both outside developers and government had drawn islanders of all ages to protest, but the underlying issue remained the overconsumption of water from Graham Lake.

BC HYDRO'S PROPOSAL

Similarly, the *Denman Rag and Bone* helped inform and rally supporters to protest a proposed underwater 500-kilovolt transmission line from Cheekye, on the BC mainland, to Dunsmuir, Vancouver Island. Because the "underwater cable is encased in a pressurized oil bath", according to Kel Kelly of the Ad Hoc Committee, the route was designed to intersect with a number of islands in order "to minimize the number of underwater splices (which pose a threat of breakage and leaking into the water)."[54] The mainland section of the line would run overland from Cheekye to Nelson Island. From there one proposal was to continue the line westward to Texada, Jedediah and then Lasqueti Island or possibly Hornby or Denman.[55] At that point it would go underwater to Dunsmuir. It would take months of study before BC Hydro and the government were able to determine the best route. In the meantime, residents from all three islands did not waste time; they provided pages and pages of evidence that among the biggest threats posed by these kinds of power lines in any location were the herbicides used to keep brush under control. The herbicide of choice, the islanders discovered to their dismay, was 2,4-D. Allegedly, the use of this herbicide on Galiano Island in 1972 had contaminated the water source, leading to two children being born with deformities.[56] On March 25, 1978, a special meeting was scheduled at Denman Island's community hall that would include speakers from Lasqueti Island and a proposal to form "a coalition of B.C. communities being adversely affected by Hydro policies."[57] Moreover, the proposed power line brought BC Hydro's entire operation, from their stated need for this power to their finances, under scrutiny.[58] Opponents confronted BC Hydro officials at public meetings.[59] People from Denman, Hornby, Lasqueti, and various districts on Vancouver Island attended these meetings to present damaging findings that questioned the very competence of the utility and the government in making these decisions. A protest action was scheduled for Parksville for April 18, 1978, and a public meeting at Courtenay for April 20.[60] Another *Denman Rag and Bone* article informed readers that "Don Lockstead, MLA for Mackenzie, . . . called for a full public enquiry, as have the Islands Trust and the Lasqueti Defence Committee."[61] Kennedy attended a BC Energy Coalition conference on Lasqueti Island, alongside fifty delegates from across the province, at which it was agreed that the local "Ad Hoc Hydro committee . . . will continue to focus on the proposed Cheekeye-Dunsmuir transmission line."[62]

Finally, in the fall of 1978, the islands learned their fate. BC Hydro had made the decision to "cross the strait directly from Texada Island to Vancouver Island thus eliminating further island hopping over Lasqueti or other islands." *Denman Rag and Bone* contributor Dave Fraser noted that

> after a year of claiming an island crossing was necessary to avoid an underwater splice, BC Hydro has reversed its position and will use an underwater splice! This was a victory of sorts for the Lasqueti Islanders and a relief to us on Denman Island.[63]

Despite their success in avoiding the use of herbicides in this case, islanders had to remain vigilant when it came to other threats to their ecosystem.

WELDWOOD ON HERBICIDES

On June 28, 1979, the Denman Island Trustees received a copy of a permit from the Pesticide Control Board granting Weldwood permission to manually spray Tordon 22K (or Picloram) on a section of its woodlot to control the growth of maple trees, which the company did not want on its land. Maples were to be felled and the stumps treated

SHARON WEAVER

to prevent sprouting.[64] Islanders had been warned of the proposal; however, the notice was so obscure that only one islander happened to find the sign. Leslie Dunsmore had been walking her dog some distance behind her house in an area without roads or obvious paths, much of which was marshland. Weldwood had already logged this quarter section, so it was not easy walking, but Dunsmore explained that she "liked bush-whacking." Despite the relative inaccessibility of the spot, she "saw a sign that was posted where nobody would ever see it, which really made me mad!" The sign informed the unlikely reader that Weldwood was soon going to "hack and squirt spray" Tordon 22K on the quarter section block of woodland to control weed species. The public were asked to report any concerns to an address provided on the sign. Dunsmore immediately wrote a letter of complaint and asked if the company knew there were about five wells off the marshland, adding that she was "concerned that the actual chemical could poison my bees and I made my living as a beekeeper."[65] In response, Dunsmore received

> a double registered letter saying I was scheduled to be heard before this panel. I later found out that I was to appear in an actual court of law. So we started looking into it and found out that Tordon 22K is the main ingredient in Agent Orange. It was me who had to present because it was an actual court of law, but there were eight of us [i.e., three members of the Community Planning Action Committee and those who prepared the case: Harlene Holm, Tom Lang, Jim Bohlen, Paul Beauchemin, and Dunsmore].[66]

Dunsmore wrote an article for the *Denman Rag and Bone*, outlining the research she and the others had conducted into the toxicity and impact on humans of Tordon 22K, the herbicide Weldwood was proposing to spray on maple stumps on Section 7 close to the Hornby Island ferry dock. Dunsmore made the obvious point that maple seeds seldom sprout in a conifer forest, as the firs shade them from the light at a crucial time in summer. Furthermore, maple seeds are wind-borne: "Thus the maple trees surrounding Section 7 will easily spread their seeds right back into the forest patch being eradicated this year."[67] The Herbicide Appeal Board heard the appeal to rescind the permit on August 14, 1979; the appellants included a Regional Board representative, three members of the Community Planning Action Committee, and Dunsmore.

"I literally stayed up and crammed the night before," Dunsmore recalled.[68] Kennedy attended the public hearing and reported on Dunsmore's presentation of her twenty-eight-page brief: "After something more than an hour, she stops. There is a split second of rapt, attentive silence and then the room explodes into prolonged applause. One senses palpable delight at having witnessed an extraordinary tour de force by a superb mind."[69] Kennedy concluded his article by noting that "widespread involvement on the herbicide issue forced Weldwood to cancel its controversial spray program." The Weldwood manager admitted that the company "had not anticipated this level of public reaction."[70] Dunsmore's presentation, Kennedy's presence at the hearing, and the applause indicate the high and effective level of community engagement with the prospect of pollution of water sources on the island. Despite differences within the community, issues related to water could elicit a great degree of agreement.

CONCLUSION

By the 1960s, the provincial government of British Columbia had recognized the vulnerability of the Gulf Islands to overdevelopment, a recognition that coincided with the arrival of the back-to-the-landers. Residents on each island covered by the Islands Trust legislation had legitimate concerns about threats to the preservation of their quality and quantity of water, whether from overlogging, improper road construction, increased tourism, small lot development, or contamination by poisonous chemicals. Indeed, the general public had become increasingly aware of the latter threat with the publication of Rachel Carson's *Silent Spring* in 1962. By a convergence of circumstances and

SHARON WEAVER

personalities, the *Denman Rag and Bone* newsletter provided a venue for such discussions at the time, linking old-time island residents and new arrivals in a grassroots exercise over a series of environmental issues. The *Denman Rag and Bone* fought the battles and recorded the victories in the struggle for local control over development plans and the quality of water resources. Like the *Denman Rag and Bone* itself, these victories were largely those of the counterculture. Many back-to-the-landers on Denman Island, as elsewhere in Canada, were recent arrivals attempting to flee the impacts of industrial capitalism—only to discover that, instead of flight, their only choice was to stand and fight.

NOTES

1   Des Kennedy, "Hack and Squirt," *Denman Rag and Bone* (hereafter *Rag and Bone*), issue 32, summer special (August 1979), 20. Tordon is a trade name for the herbicide Picloram.

2   Richard H. Grove, *Green Imperialism: Colonial Expansion, Tropical Island Edens and the Origins of Environmentalism, 1600–1860* (New York: Cambridge University Press, 1995).

3   While conceivably a bridge could be built between Denman and Vancouver Island (or the mainland, as Denman Islanders call it), the small island's remoteness and the relatively small population on Vancouver Island, where one end of the bridge would be located, keep Denman as a small maritime island ecology with all of its vulnerabilities.

4   *Gulf Islands Ecosystem Community Atlas* (Vancouver: Canadian Parks and Wilderness Society—British Columbia Chapter/ Parks Canada, 2005), 8, accessed 6 September 2012, http://cpaws. org/uploads/pubs/atlas-gulf.pdf.

5   Ibid., 30.

6   Jim Bohlen, author of *The New Pioneer's Handbook: Getting Back to the Land in an Energy-Scarce World* (New York: Schoken, 1975) and a founding member of Greenpeace, moved to Denman Island in 1974. Some consider Bohlen and his wife, Marie, to be the island's most famous environmentalists; while that may have been the case, they were not involved with the *Rag and Bone*, nor were they mentioned during the thirty interviews I conducted on the island. This may have been due to their macro environmental focus, which differed vastly from the micro focus of the majority of back-to-the-landers on

the island. (This chapter is from a larger study of the back-to-the-land movement on Cape Breton, Denman, Hornby, and Lasqueti islands.)

7   Linda Adams, "In Depth," on Islands Trust website, accessed 29 January 2011, http://www.islandstrust.bc.ca/tc/pdf/indepthcaoita.pdf.

8   Robert L. Bish, *Local Government in British Columbia* (Richmond, BC: Union of British Columbia Municipalities in co-operation with University of Victoria School of Public Administration, 1987), 33.

9   Ibid., 60. The source for population figures is the 1981 Census of Canada. Not only was there a lack of island representation on the regional boards, but some islands were assigned to regional districts that made little geographical sense. Lasqueti Island, with a population of 316 permanent residents, was part of the Powell River Regional District (population 19,364), which was far to the east on the mainland with no transportation links to Lasqueti. It would have made far more sense for Lasqueti to have been in the same regional district as Denman and Hornby islands, with which it has social and cultural similarities.

10   Bish, *Local Government in British Columbia*, 59.

11   Ibid., 46.

12   Ibid., 59.

13   Ibid., 59–60.

14   Ibid., 60–61.

15   According to census data, the population of Denman Island was 250 permanent residents in 1971 and 589 permanent residents in 1981.

16   Des Kennedy, interview with the author, Denman Island, 26 January 2008.

17   Sandy Kennedy, *Rag and Bone*, issue 18; mislabeled as 17 (July 1976), 7. Sandy Kennedy worked for Statistics Canada as a census taker; she gives this figure as the approximate number of permanent residents on the island. The number of permanent households was approximately 167, and vacation homes approximately 55. Denman Islanders often compare the size of their island, at 51 square kilometres, to that of Manhattan (87 square kilometres). Of course, this comparison is not useful for ecological purposes, because Manhattan sits in the Hudson River and does not have the same freshwater limitations.

18   Editor, "Dear Reader," *Rag and Bone*, issue 25 (February 1978), 2.

19   "Bits & Pieces" or "Bits and Pieces"—the title format sometimes changed within the same issue.

20   Kennedy, interview.

21   Ibid.

22   Manfred Rupp, interview with the author, Denman Island, 27 November 2007.

23   Ibid. It was a common practice during this period for people with a desire to own a piece of land to purchase property

as a group; in this case, eight individuals living in Vancouver (four couples) searched for land together.

24 Bish, *Local Government in British Columbia*, 40.

25 Ronald Inglehart, *The Silent Revolution* (Princeton: Princeton University Press, 1977), 3-18. For further insight see Kathryn Harrison, "Environmental Protection in British Columbia: Postmaterial Values, Organized Interests, and Party Politics" in *Politics, Policy, and Government in British Columbia*, ed. R. K. Carty (Vancouver: UBC Press, 1996), 290-311.

26 D. Kennedy, "The Islands Trust Act," *Rag and Bone*, issue 1 (May 1974), 12.

27 "The Islands Trust is a unique form of government, created by the province in 1974 to control unbridled development and to 'preserve and protect' the islands." Islands Trust Act, S.B.C. (1974), C43, accessed 19 September 2015, http://www.islandstrust.bc.ca/ trust-council/islands-trust-act/ islands-trust-legislative-history. aspx.

28 "Editorial," *Rag and Bone*, issue 2 (June 1974), 2.

29 Manfred Rupp, "Postscript," *Rag and Bone*, issue 2 (June 1974), 8.

30 W. A. Hopwood quoted in Manfred H. Rupp, "Commentary," *Rag and Bone*, issue 3 (July 1974), 8.

31 Rupp, "Commentary," 9.

32 Des Kennedy, "Editorial," *Rag and Bone*, issue 11 (June 1975), 2-3.

33 Walton was not a back-to-the-lander, nor an original settler, but one of the retirees who had moved to the island permanently.

34 Peter McGuigan, "Speakers' Corner," *Rag and Bone*, issue 4 (August 1974), 4.

35 Daphne Morrison, "Ratepayers' Report," *Rag and Bone*, issue 4 (August 1974), 12.

36 Manfred Rupp, "Roads, Roads, Roads . . .," *Rag and Bone*, issue 6 (November 1974), 3.

37 Bish, *Local Government in British Columbia*, 89.

38 Manfred Rupp and Daphne Morrison, "The Denman Bump and Grind," *Rag and Bone*, issue 8 (January 1975), 8.

39 Des Kennedy, "Editorial," *Rag and Bone*, issue 16 (March 1976), 3. The Ratepayers' Association is frequently referred to simply as "Ratepayers."

40 Ibid.

41 Beverly Meyer and Des Kennedy, "Rumours Galore," *Rag and Bone*, issue 12 (Summer 1975), 18.

42 Des Kennedy, "Editorial," *Rag and Bone*, issue 18; mislabeled as 17 (July 1976), 2.

43 Kirk, quoted in Kennedy, "Editorial," (July 1976), 2.

44 Kennedy, "Editorial," (July 1976), 3.

45 As property taxes rose for all islanders, it can be argued that

summer residents obviously had sufficient income for a second residence, whereas islanders were limited in their ability to generate greater income year after year.

46   Paul Bailey, "Bits and Pieces," *Rag and Bone*, issue 31 (May 1979), 3.

47   Bish notes that "for the unincorporated areas of regional districts, the approving officer is an employee of the Ministry of Highways and Transportation" (*Local Government in British Columbia*, 112).

48   "Editorial," *Rag and Bone*, issue 2 (June 1974), 6.

49   Bailey, "Bits and Pieces," 3–4.

50   Ibid.

51   Kennedy noted in an earlier edition of the *Rag and Bone* that "approval of subdivisions still rested with the Department of Highways; planning and zoning remained under Regional Board control. The trust participated in these matters on an advisory basis. It quickly became evident that Regional Boards resented Trust interference in what had been their sole domain" ("Bedtime Reading," *Rag and Bone*, issue 22 [May 1977], 16). This could explain the apparent about-face by the department of highways.

52   Kel Kelly, "By-Law Burial," *Rag and Bone*, issue 32, summer special (August 1979), 2–3.

53   Des Kennedy, "Check the Malls for Scowls & Folly," *Rag and Bone*, issue 33 (Christmas 1979), 3.

54   Kel Kelly, "an electrifying experience," *Rag and Bone*, issue 25 (February 1978), 17.

55   Des Kennedy, "an electrifying experience," *Rag and Bone*, issue 25 (February 1978), 15.

56   Ibid.

57   Kelly, "an electrifying experience," 17.

58   Ibid.

59   Des Kennedy, "I Protest!" *Rag and Bone*, issue 27 (July 1978), 10.

60   "Ace Rolls On," *Rag and Bone*, issue 26 (April 1978), 4.

61   Robbie Newton, "Hydro," *Rag and Bone*, issue 26 (April 1978), 15.

62   Des Kennedy, "Bits and Pieces," *Rag and Bone*, issue 28 (October 1978), 19.

63   D. Fraser, "Blinded by the Light," *Rag and Bone*, issue 29, A Year in the Life of an Island: Special Edition (January 1979), 26.

64   Leslie Dunsmore, "Hack and Squirt," *Rag and Bone*, issue 32, summer special (August 1979), 18.

65   Leslie Dunsmore, interview with the author, 11 December 2007.

66   Ibid.

67   Dunsmore, "Hack and Squirt," 19.

68   Dunsmore, interview.

69   Kennedy, "Hack and Squirt," 20.

70   Kennedy, "Check the Malls for Scowls & Folly," 3.

3

"Good Ecology Is Good Economics": The Slocan Valley Community Forest Management Project, 1973–1979

Nancy Janovicek

Scholarship on the contemporary environmental movement emphasizes conflict. The "War in the Woods" in British Columbia in the
1990s evokes images of confrontations between grandmothers and
police, young hippies and loggers, and First Nations peoples and government officials. Disillusioned by the unwillingness of governments
to implement policy that recognizes the urgency of the rapid depletion
of the earth's natural resources and the interconnectedness of ecological issues and social inequalities, many environmental activists and
scholars have rejected the politics of compromise and coalition. Some
believe that green democracy can only be achieved through autonomy from the state and in conflict with local stakeholders who make
their living in the woods. As political scientist Laurie Adkin argues,
"It is time to set aside the master's tools of 'sustainable development'
and turn our efforts toward the realization of ecological democracy."[1]

Ecological justice, in this framework, is not compatible with economic development.

In contrast to this depiction of the divisiveness of environmental politics, this chapter examines a community-based research project conducted in the West Kootenays region of British Columbia in the 1970s that sought to bridge the divisions within the community. This project aimed to develop an ecologically sound land-use plan that accommodated the economic and political interests of environmentalists, loggers, recreationists, trappers, and farmers. The Slocan Valley Community Forest Management Project (SVCFMP), a project funded by the federal Local Employment Assistance Program (LEAP), began as a feasibility study to create new employment opportunities in the forest that did not harm the environment. Researching and writing the report was a deeply democratic process. The "new homesteaders," who started to migrate to the Slocan Valley in the mid-1960s as part of the back-to-the-land movement, initiated the study, but its appeal to the residents of the valley rested on the coalitions that they built with people who had lived there for generations. As Kathleen Rodgers has shown, back-to-the-landers introduced new political theories and practices to the area.[2] I argue that they also built on the local political culture established by the Doukhobors, unions, and old left politics. Economic vulnerability, common in resource-based economies, and a sense of rural alienation from senior levels of government, which was ingrained in the West Kootenays, informed these politics. New homesteaders lived according to ideas associated with the counterculture, such as local control over resources and government and a "DIY" approach to daily life.[3] These values resonated with the western Canadian co-operative tradition and the pacifist and communal beliefs of the Doukhobors. Most important, all people who wanted to build a life in the Slocan Valley agreed that government policy should ensure that valley residents be the primary beneficiaries of economic development and that their children enjoy the benefits of local resources. This applied to both the timber and the beauty of the woods.

The SVCFMP's 1975 report to government officials and universities combined environmental stewardship and the protection of forestry jobs as the foundation for economic development. The authors of the report argued that responsible forest management should be based on the "sustained yield of all resources, from fish to water to trees."[4] Their proposal insisted that local economic independence could be achieved only with the complete integration of all of the valley's resources: timber, agriculture, fish, furs, and water. Only by decreasing dependence on logging could the community ensure economic stability. A second edition of the report, published for community stakeholders the following year, added a proposal to conserve the Valhalla Range as a provincial park. The premise of the research was that "good ecology is good economics."[5] This aspect of the forestry project supports recent research demonstrating that the tendency to pit environmentalists against workers does not capture the complexities of the history of the environmental movement.[6] This chapter also examines the relationship between local organizing and the state in countercultural and environmental politics. The federal and provincial initiatives in the early 1970s that promoted local civic engagement as a means of solving social and economic problems gave this community-based project political credibility. Those who were involved in the local consultations to produce the report did not reject government involvement in the development and implementation of economic policy; rather, they insisted that the government follow the direction of local people because they were the most knowledgeable resource managers.

The radical environmental politics of the final report of the SVCFMP, which scholars attribute to the new ideas that came from the back-to-the-land movement, has received cursory attention in studies about British Columbia wilderness politics that focus on conflict.[7] My analysis takes a different approach. I argue that the radical environmental politics that the report proposed drew from existing economic development plans that emphasized local control over resource management. The new homesteaders respected the knowledge and experience of people who had worked in the woods for generations and collaborated with them to develop a forest management proposal intended to address the economic and cultural goals of all constituencies in the valley. Like the activists who led the campaign to stop herbicide spraying on Denman Island that Sharon Weaver discusses in this volume, the SVCFMP sought to mobilize all valley residents. The Valley Resource Society became a coalition among groups with disparate political goals, but it emerged from conflict. I begin by examining the development of the coalition between the back-to-the-land community and long-term residents who were upset by forestry practices that threatened the future economic viability of the valley. A central argument of the SVCFMP report was that single-use resource management, focused on "timber mining" for foreign profit, had destroyed local businesses that had managed the woods in an ecologically sound manner. Rapid exploitation of a single resource had created a precarious economy. Economic planning, the report argued, had to follow "enforce[d] guidelines based on Nature's ability to regenerate" in order to prevent the destruction of "both our forests and the Slocan Valley community it supports."[8] The SVCFMP recommended an alternative model to wasteful logging practices that accommodated the different needs of loggers, farmers, recreationalists, and trappers.

Moreover, the proposed integrated resource model insisted on protecting the Valhalla Range, an area of the Selkirk Mountains that was treasured for its old-growth forests and spectacular summits. Pre-contact artifacts of the Salish-speaking Sinixt First Nation and the remnants of early twentieth-century logging equipment make this area an important cultural heritage site. This wilderness was also the habitat of endangered species. Thus, conservation became a core principle of economic development in the Slocan Valley.

BUILDING A COALITION

Between 1966 and 1971, the population of the Slocan Valley increased by 420 persons, a trend that reversed years of outmigration. According to the SVCFMP final report published in 1976, 225 young families had arrived in the valley since 1970, comprising about 15 per cent of the population.[9] Drawn to the Kootenays by cheap land, these families moved to the country to get away from the rising cost of living and pollution in the cities. They hoped that by growing their own food and living a simpler life they would be self-sufficient and that they would be able to develop local alternative economic models based on mutual aid. There were angry, and sometimes violent, clashes between the newcomers and long-time valley residents who opposed their values. The final report acknowledged these hostilities, but observed that "deeper than these feelings, however, is a unifying and commonly shared feeling of love for the Valley landscape, its hills, waters, wildlife."[10] Back-to-the-landers and people who had lived in the valley for generations learned that they shared a commitment to locally controlled economic development that was attentive to the environmental impact of a resource-based economy.

At the individual level, mutually co-operative relationships developed between new homesteaders and their neighbours, especially with Doukhobors.[11] These friendships did not lead to the general acceptance of the back-to-the-landers, though. The people who moved to the Slocan Valley from cities in the United States and Canada introduced political ideas, family forms, and lifestyles to the Slocan Valley that challenged the area's predominantly conservative social values. Despite the hippies' efforts to co-operate with the community, a small and vocal group insisted that these newcomers were an immoral influence on the area. They were angry at the influx of young people who lived alternative lifestyles, used illegal drugs, and practiced public nudity. Further, many back-to-the-landers worked part-time or seasonal jobs and therefore relied on unemployment insurance. This angered some long-time residents, who believed that they were abusing social welfare programs to avoid work. Back-to-the-landers often referred to these unhappy neighbours as "the Anglos"—or, more pejoratively, "the rednecks"—to distinguish them from the supportive Doukhobors who taught them rural and farming skills.

Federal and provincial initiatives that encouraged civic engagement helped the back-to-the-landers establish programs and institutions based on their values.[12] Those who resented the influx of Americans and hippies in the valley viewed their use of such programs as another abuse of the system. In 1973, a long letter to the editor of the *Nelson Daily News*, signed by thirty residents, described the newcomers as "freeloaders" who had "no intent in proving that they were willing to live in harmony with us 'native people' of this valley" and, further upsetting the residents, had begun "receiving grants in astonishing amounts for some of the most ridiculous projects imaginable."[13] Local lumberman Don Sutherland organized valley residents to "stop the grants, unemployment, nudism, drugs and unfair law enforcement."[14] This group opposed the newcomers' bid to join the Civic Action Committee, a provincial initiative intended to encourage local governance, and changed the mandate of the committee to focus on removing the "hippie element" from the valley.

Sutherland, a member of the International Woodworkers of America, had initially supported the application for research into alternative economic models for the valley's forestry industry. He withdrew his support because he was outraged that the newcomers were receiving government funding for projects that he deemed to be of little value. When Michael Pratt, a member of the SVCFMP, went to collect literature about the community forestry project that Sutherland had agreed to distribute, Sutherland punched Pratt and threatened him with a club because he refused to leave without the documents. Pratt was not a "typical hippie." A Canadian who had emigrated from England as a child, he was forty-one at the time of the altercation with Sutherland. Pratt was a father of four, held a PhD in natural biology, and had left a government job in Vancouver to move with his family to the valley for health reasons. His children went to the Free School, and this association with the local counterculture may have compelled Sutherland to call Pratt a "filthy stinking hippy" and to push him when he refused to leave.[15] Because of the increasing tensions concerning federal funding for projects sponsored by the newcomers, the SVCFMP decided to delay applying for LEAP funding to support its research until the next year.

Conflicts between the "hippie sect" and "the Valley natives" created deep divisions in the community, but the SVCFMP persisted and managed to attract people from the Doukhobor and "Anglo" communities who supported the idea of local management of the forestry

3.1 Slocan Valley Community Forest Management Project Steering Committee. Source: *SVCFMP Final Report*, 2nd ed., back cover.

industry. In 1974, the group received a fifty-thousand-dollar LEAP grant to raise community awareness about how government policy on economic development affected the lives of area residents and to promote economically and ecologically sustainable forest management. The original SVCFMP committee had included twelve people, but it grew as people learned about the project. As a photo of the steering committee (figure 3.1) subtly underlines, the committee included representatives from the three key communities in the valley: Sam Verigin and Peter Bloodoff Jr. were Doukhobor; John Braun was a local woodsman, hunter, and trapper who joined the group along with his friend Jim Warner (not shown), a millworker; Frank Nixon was a farmer and sawmill worker; Tell Schrieber, M. L. Thomson, Bob Ploss, and Conrad (Corky) Evans were new to the valley. The committee hired Evans to be the administrator for the project. Evans recalled that he was selected over applicants with PhDs because of "people like Jim Warner and Johnnie [Braun]. . . . I was logging, and they could relate to that. So I had an urban life . . . [and] I had work experience that they could relate to. So I was kind of a compromise."[16] All of the employees of the project were young adults as stipulated by the LEAP program, but Evans also observed that the wages allowed by the federal job-creation program were too low to attract professional people.[17]

The project collected questionnaires from people in the valley as well as letters from people interested in the project. This enthusiasm for an alternative to the New York–based Triangle-Pacific (Tri-Pac), which controlled 90 percent of the logging tenure in the area, certainly reflected concerns about high unemployment in the region.[18] Federal and provincial promotion of local citizen engagement—albeit for political reasons that may not have accorded with the goals of valley residents—also shaped the debates about who should control the local industry. Provincially, the New Democratic Party (NDP) government that came into power in September 1972, under the leadership of Dave Barrett, supported local involvement in policy decisions. The goal of the Community Resource Boards Act, passed in 1974, was to empower citizens to identify social problems and use their knowledge of local needs to develop solutions and services.[19] The provincial government's encouragement of local political engagement extended to land policy. The Agricultural Land Commission established in 1973, held public forums to determine which land should be protected from development to sustain the agricultural sector. Although these hearings were controversial and caused divisions at the local level, they demonstrated a commitment to decentralized decision making.[20] In his analysis of forestry policy and the environmental movement in British Columbia, Jeremy Wilson explains that the Barrett government introduced policies that challenged the government-industry pact and insisted that British Columbians deserved a larger share of their resources. For instance, Bob Williams, the minister of lands, forests, and water resources, did not support the sustained yield policy of previous governments. This central tenet of forestry policy, which dated back to the Sloan Royal Commission in 1945, held that

NANCY JANOVICEK

old-growth forests were a rotting resource that needed to be harvested and replaced with scientifically managed tree farms. Ultimately, the NDP government did not introduce radical environmental policy. But Williams's support for diversified control of the industry, as well as his belief that talented laypeople were best equipped to manage their resources, created political spaces where citizens could present alternatives to government policy that was not attentive to local needs.[21]

Those who joined the SVCFMP brought a range of experience and educational expertise to the research, but most of them had no formal training in silviculture. Their vision for ecological forest management rested on their experience in the logging industry and their anxiety about the negative impact of increased mechanization, introduced in the 1960s, not only on the logging industry but also on other businesses, especially agriculture, trapping, mining, and recreation. Both long-term and new residents witnessed waste, human-caused flooding, and the destruction of wildlife habitat caused by clear-cutting and poorly planned access routes.

They also brought different views about environmentally sensitive economic development, ranging from deep ecology to pragmatic conservation. Evans recalled that he built bridges among people by asking them to role-play the different constituencies in the community. Drawing on his background in community theatre, Evans challenged them to defend interests of a group to which they did not belong so that they would learn to understand other people's positions. He explained why this was a productive method:

> If you're in a room with a bunch of people, it's better if they're kind of actors than if real loggers—because if they're real loggers and miners they're terrified, right, that "you're going to hurt me." . . . But if you're acting for the miners, then you go, "there's a bunch of silver here. What do you mean you want to make this into a park?" . . . And there's no fistfights because you're articulating a position, which everybody can see is real. . . . The fact that Peter Bloodoff Sr. and Johnnie and Jim and I don't remember who else

was in the room—it had a big moderating influence on our impetuousness, or youth or whatever *you* were trying to make happen. You knew that on Saturday it was going to have to be saleable to everyone who had been trying to do it all their life. [If] you said trapping was evil, as some urban people are likely to do, you're going to have to say it to Johnnie, you know.[22]

The final report, written by Dan Armstrong and Bonnie Evans, explained why this process was effective. In the group's first meetings, members had been assigned responsibilities based on their personal expertise and interests. However, they realized that this process prevented people from compromising with those who held different views: "*Placing labels on people*, such as 'the economist' or 'the conservationist' defines and therefore limits their involvement in the problem."[23] Corky Evans's method compelled people to defend disparate positions, helping them to understand the interrelationships between different economic sectors and the natural environment.

Working in the same room also meant those who had recently moved to the valley learned to appreciate the experience and views of their elders. They had to compromise. Using data from the Canadian Land Inventory, a project that produced aerial photographs of the country, members of the community gathered around a large table with a sheet of vellum placed over a base map of land forms and traced the different types of land use onto the map. Evans explained that this allowed the group to "figure out where things could happen with less conflict or what things shouldn't happen."[24] Most important, all of the people working on the project lived in the valley. Their investment in creating a viable alternative to single resource management that prioritized logging over other industries helped to overcome some, though not all, of the disputes over land use in the valley. The final report endorsed economic diversification and defended the protection of less profitable businesses, such as agriculture and trapping, by explaining how different industries and cultural groups had historically worked together in the Slocan Valley.

PROPOSING AN ECOLOGICAL ALTERNATIVE

The fundamental argument of the report was that decentralized control over the logging industry was the only responsible way to manage the valley's timber resources. Unlike multinational companies, which were primarily concerned with increasing their profit margins, the community had an investment in the continued viability of all resources. The report recommended an economic model based on integrated resource management that recognized the changing interrelationships among different resources. The proposed economic plan made space for local businesses and independent loggers who had been pushed aside by Tri-Pac's virtual monopoly. It also defended "non-tangible" resources—in particular, the bucolic scenery in the Kootenays—which were becoming increasingly valuable as more people developed an environmental consciousness.

The argument for integrated resource management was grounded in history. Evans recalled that when the committee began to discuss community-based land-use policy, they followed industry models that focused on economic use. Peter Bloodoff Jr. intervened to suggest that they begin by examining the nature of the landscape and the history of the community. Ultimately, the report reflected Bloodoff's suggestion: it opens with a discussion of the natural history of the valley and explains the development of both climax forest, characterized by the achievement of a stable ecosystem of self-sustaining forests, and successional forests, which establish themselves when floods or fires disrupt the closed system of the former. The arrival of humans caused the "seemingly backward evolution" that had resulted in the predominance of successional forests over climax forests, especially in the previous one hundred years.[25] The authors then discussed the different groups of people who had lived in the valley, including the Sinixt, Anglo-Saxon homesteaders and miners, Doukhobors, interned Japanese Canadians during World War II, and back-to-the-landers. The histories of these groups demonstrated how subsistence and local market farming, logging, and mining were historically interconnected. A key criticism levelled in the report was that the emphasis on

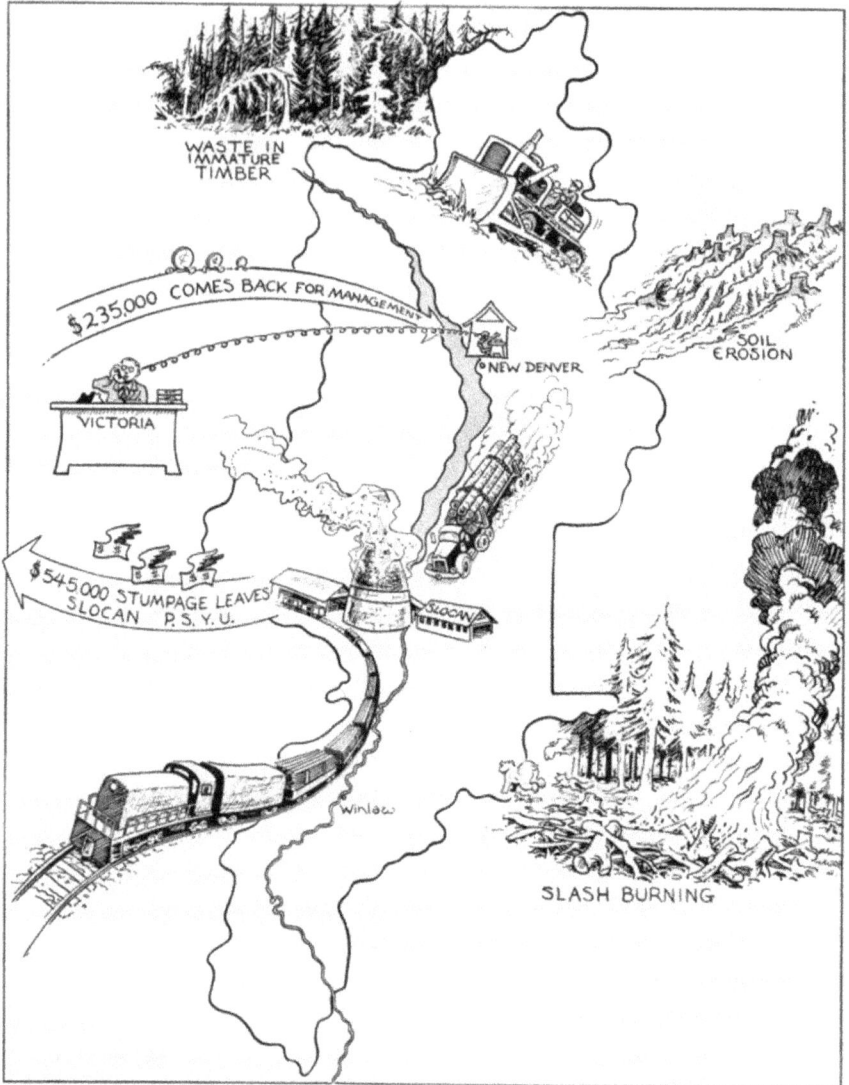

Inside the illustration:

WASTE IN IMMATURE TIMBER

$235,000 COMES BACK FOR MANAGEMENT

SOIL EROSION

NEW DENVER

VICTORIA

$545,000 STUMPAGE LEAVES SLOCAN P.S.Y.U.

SLOCAN

Winlaw

SLASH BURNING

THIS IS SINGLE RESOURCE MANAGEMENT--

Forest management is geared for timber extraction at the expense of our other resources. Short-sighted policy drains the community of its natural bounty. We see waste, environmental damage, and a limited return of management funds for resource agencies in Nelson and New Denver.

3.2 George Metzger, "Single Resource Management." Source: *SVCFMP Final Report*, 2nd ed., pp. 3–98.

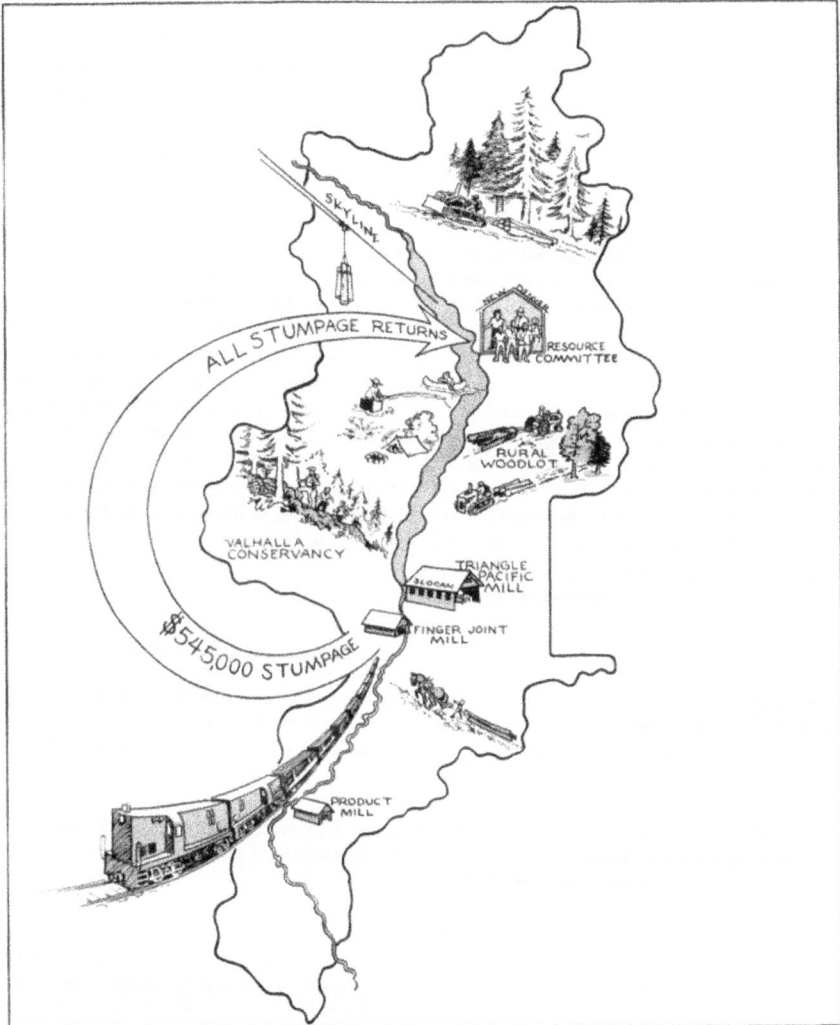

THIS IS INTEGRATED RESOURCE MANAGEMENT—

Recommendations based on long-term considerations of all our resources have been implemented. Forest management proceeds on a true "sustained yield" basis, while the needs of the local community and its surrounding environment are met in an ecologically sound way.

3.3 George Metzger, "Integrated Resource Management." Source: *SVCFMP Final Report*, 2nd ed., p. G-4.

logging as the region's economic stimulus destroyed good agricultural land. Protecting farms was important because arable land was restricted to the valleys along the rivers and lakes. Even though agriculture did not drive the local economy, it was vital to many family economies.[26]

This historical overview laid the foundation for the analysis of the inadequacy of single resource management and defended the proposal for integrated resource management. The differences between "the existing situation" and the "proposed situation" are captured in George Metzger's illustrations (figures 3.2 and 3.3).[27] Damaging logging practices, such as clear-cutting and slash-burning, destroyed watersheds. Wasteful logging also threatened the continued viability of logging because it destroyed the conditions that would make forest rejuvenation possible. Moreover, government management plans did not consider the importance of resources besides wood fibre, and they ignored the negative impact of irresponsible logging on agriculture, recreational land use, trapping, and fishing.

The report was equally critical of the consolidation of small, locally owned logging mills in the hands of foreign-owned multinationals. Increased harvests under these larger companies improved economic stability for most people in the valley, but "the virtual exclusion of the independent small operator from any forest activity has had ecological and sociological effects that have been unnecessary and damaging."[28] Logging happened on publicly owned Crown land, which should have instilled a sense of community responsibility for the resource. Instead, people considered "the forest as an economic extension of 'the company,' and not as their own environment, and therefore see little reason why they should worry about 'the company's trees.'"[29] Foreign ownership and government policy also meant that all of the profits from the region's key resource left the valley. In comparison to other North American jurisdictions, stumpage fees in British Columbia were very low. Most of this money went to the provincial coffers, while the funds that returned to the Slocan Valley went to the logging company's managers, who did not have a stake in the future vitality of the region. The SVCFMP criticized modern "efficient" logging practices that wasted most of the wood as well as the failure to produce diversified secondary industries. The report emphasized the urgent need for change: "We are sitting on one of the largest and most varied, as well as the last great North American, forests. We are cutting it as fast as we can, with little thought to the future, selling it too cheaply, in semi-furnished form, or wasting it, and then paying exorbitant prices (plus duty) for the manufactured items other nations make out of our wood."[30] The timber resources from the West Kootenays were producing secure employment in other nations, and this increased dependence on external markets created job insecurity at home.

Only local control over the forest could reverse these destructive practices. The SVCFMP argued that the valley's lumber resources were overcommitted ecologically and demanded a reduction of the allowable annual cut. Referring to a 1955 report by Ray Gill on logging in the Slocan Valley, the group argued that the government had long been aware of the negative impact of logging on the region. They recommended the implementation of Gill's recommendations, which had pointed to the need for selective logging in sensitive areas.[31] Integrated resource management was the only ecological alternative. The community was an important resource, too, because it "possesses an attribute that is often overlooked in forest management, that of permanence."[32] In contrast to the bureaucrats in the provincial capital, Victoria, and the owners of the logging companies, residents of the community would have to live with the consequences of good and bad policy. The report recommended the establishment of a resource management committee made up of local residents. In public hearings to promote the findings, Evans insisted that local residents would be less likely to exploit and destroy land than international companies, who were not invested in the community.[33]

The committee rejected forestry management that viewed the woods as a "boundless source of timber" and called for policy that "allow[ed] the nature of our resources themselves to dictate their utilization."[34] To diversify and decentralize the forestry industry, the report suggested a system of rural woodlots, ranging from 10 to

1,500 acres, to help supplement the incomes of people who farmed. Farming—a "common denominator" for most of the people living in the valley—needed to be preserved "in these days of escalating food prices and debate on the nutritional value of many retail foods."[35] The rural woodlot system would also protect ecologically sensitive areas by adopting older technologies such as horse logging. The development of mills to produce other materials, such as wood chips and cedar shingles, would ensure that the entire tree was used, thus reducing waste.

An interesting aspect of this report was its insistence on the maintenance of the diversity of wildlife as a key component of the integrated management of resources. The protection of this habitat was not restricted to conservation areas, but also included places where responsible logging could occur. The report identified "critical wildlife areas," which had been "compiled by local knowledge and observation over many years," that needed special protection.[36] Local woodsmen and naturalists were concerned that logging activity had depleted the population of mule deer, white-tailed deer, mountain goats, and caribou, as well as fish stocks. Even though the local grizzly bear population was healthy, logging was posing a serious threat to its habitat. Protecting grizzly habitat was necessary because those bears were less adaptable than black bears; this was a key reason for the proposal to preserve the Valhalla Range as a conservancy area. Local recognition of the need to preserve wildlife habitat reflected a shift in government conservation policy after 1970, which asserted that saving endangered species depended on the conservation of the places where they lived.[37]

Conservation of the Valhalla Range did not comfortably comply with the key goal of ensuring that people could make their livelihood in the woods. The committee explained that "parks, in general, are certainly inconsistent with our vision of land use, but so is Victoria, so we decided to play it safe and protect this very special place."[38] Adding the proposal for a conservancy area demonstrated that a significant number of valley residents believed old-growth forest should be regarded as "a sanctuary, a museum of and a monument to the natural history and scenic beauty of the region."[39] Unlike earlier proposals for a park, which had come from people who did not live in the valley, this report accommodated the traplines that currently existed in the proposed conservation area because furs were a renewable resource. It also supported properly managed sport hunting.[40] However, this accommodation of older valley lifestyles was unacceptable to activists who had joined the SVCFMP to conserve the Valhalla Range.[41] Many of them left the Valley Resource Society in 1975 to form the Valhalla Wilderness Society, to focus on banning logging in the range and to lobby for the creation of a park (which they achieved eight years later).

One of the most important recommendations for the committee—and most likely the reason why politicians did not endorse the report—was that a local resource committee should have control over the annual $545,000 in stumpage fees. In an interview with Jeremy Wilson, Bob Williams praised the work of the SVCFMP: "I still think it is probably the finest social economic analysis in modern history in British Columbia. . . . So I was impressed with it. But I was still a pragmatic politician, saying 'How far can we go?' We were talking about the Crown jewels and all those ragamuffins up in this nowhere, beatnik valley want the jewels."[42] Evans recalled that when they presented the report to Williams and asked to manage the timber he responded, "Not on your life!" The SVCFMP rejected a counteroffer for a locally owned sawmill that would have created twenty jobs. The delegates insisted that until the annual allowable cut was reduced, the future of the industry would not be protected. When the NDP lost power, Williams invited Evans to present the SVCFMP report to a class that Williams was teaching at Simon Fraser University. Evans agreed to speak to the class—on the condition that Williams come to the Appledale Community Hall and apologize to the community for rejecting the report. He did.[43]

After their unsuccessful lobbying of the provincial government, which advised them that they were "'pretty naïve' if [they] thought that [they] could control [their] own destiny," the committee formed the Valley Resource Society to continue the discussions about land-use policy and resource management.[44] Members of the community debated the recommendations of the report at public meetings held throughout the valley that involved government officials and representatives from Tri-Pac. These meetings were well attended; according to media reports, 40 to 150 people showed up to discuss the plan.[45] Activists advised people to take control over public land, which was owned by the residents of the Slocan Valley, not the government or the logging multinationals. An article in the *Arrow*, an alternative newspaper published in Castlegar by back-to-the-landers, insisted that all valley residents needed to work together to implement the recommendations of the report: "And know your allies. Most of the people involved in the Forest Industry, from fallers to Foresters to Government workers to Educators are *good* people. Looking for bad guys is dissipation of energy."[46] The Slocan Valley elected members to serve on the resource management committee, but they were not successful in gaining control over forestry. And they could not prevent the massive layoffs in the industry in the 1980s when the Kootenay Forest Products sawmill closed.

CONCLUSION

Scholarship on countercultural communities of the 1960s and 1970s tends to focus on their rejection of social conventions and advocacy of lifestyles outside of the mainstream.[47] Similarly, the history of the environmental movement makes a clear distinction between the conservation movements in the early and mid-twentieth centuries and the contemporary environmental movement. The social movements of the 1960s certainly changed political engagement, but there is much to learn from the continuities between the new radicalism and older forms of political protest. In the West Kootenays, many of the back-to-the-landers came to respect the knowledge of the elders in the community. This story reminds us that in terms of environmental politics, it is important to examine the contributions of people who work in the woods. The SVCFMP represents a moment when people with diverse political positions worked together to try to protect the valley for future generations. This process of political engagement also taught them to find allies in their neighbours and to recognize

NANCY JANOVICEK

that these people were their best teachers. This is perhaps best illustrated in the report's dedication, which reads, "to John Braun and Jim Warner and others like them. If we did a good job, it's because of your vision. If we didn't, it isn't as if you didn't try."[48]

The democratic processes that the Valley Resource Management Society demanded did not lead to the implementation of co-operative policy development in the 1970s. But it was a precursor of subsequent provincial government policy. In an effort to bring peace to the War in the Woods that defined BC environmental politics in the 1990s, Mike Harcourt's NDP government established the Commission on Resources and Environment (CORE) in 1992. CORE's mandate was to facilitate community involvement in regional planning in order to advise the government on land-use policy and environmental regulation. The commission instigated planning groups in four regions, including West Kootenay–Boundary Table.[49] Basing government policy on collaboration with community stakeholders was a strategy designed to build consensus among government, industry, labour, First Nations peoples, and environmentalists. In his analysis of CORE's deliberations in the Slocan Valley, Darren Bardati argues that despite the government's commitment to engaging local residents in policy making, community and government were not on a level playing field. As a result, local residents felt betrayed by a process that was not able to break the industry's control over forest management plans.[50] In part due to criticisms of the CORE process, the NDP government later implemented smaller-scale Land and Resource Management Planning (LRMP) consultations with regional groups to make recommendations on land-use policy, a process that Wilson calls "hyperconsultative."[51] The LRMP process was more successful. The Kootenay-Boundary Resource Management Plan, tabled in 1995, recognized agriculture and ranching as important industries and laid out a strategy to integrate forest management and agricultural land-use policy.[52]

These provincial consultations have carved out spaces for community forests. Today, logging co-ops coexist with multinational companies in the West Kootenays. One example is the Harrop-Procter

Community Co-op, which runs a community forest located on the south shore of the west arm of Kootenay Lake, twenty-five kilometres east of Nelson. Founded in 1999, the co-op has developed ecologically sustainable logging methods and is committed to providing local people with "socially and economically equitable" jobs.[53] Their land-use strategy draws on ecosystem-based plans to develop a diversified economic foundation that also incorporates agricultural development and ecotourism, a land-use strategy that echoes the goals of the Valley Resource Society. Clear-cutting continues, and neighbours are still divided on how to manage the region's resources, but these co-operative models prove that sustainable forestry is viable and that good ecology is indeed good economics.

NOTES

1   Research for this chapter was supported by the Social Sciences and Research Humanities Council of Canada. Laurie E. Adkin, "Preface," in *Environmental Conflict and Democracy in Canada*, ed. Laurie E. Adkin (Vancouver: UBC Press, 2009), xii.

2   See her contribution to this volume and her longer study, *Welcome to Resisterville: American Dissidents in British Columbia* (Vancouver: UBC Press, 2014).

3   Peter Braunstein and Michael William Doyle, eds., *Imagine Nation: The American Counterculture of the 1960s and '70s* (New York: Routledge, 2002).

4   Valley Resource Society, *Slocan Valley Community Forest Management Project Final Report*, 2nd ed. (Winlaw, BC: Slocan Valley Resource Society, 1976),

xiii (hereafter *SVCFMP Final Report*).

5   Ibid., xi.

6   Eryk Martin, "Class Politics, the Communist Left, and the (Re) Shaping of the Environmental Movement in BC, 1973–1978" (paper presented at the Canadian Historical Association Annual Meeting, Montreal, 30 May, 2010). For a discussion of the division between environmentalists and workers, see Richard White, "'Are You an Environmentalist or Do You Work for a Living': Work and Nature," in *Uncommon Ground: Rethinking the Human Place in Nature*, ed. William Cronon (New York: W. W. Norton, 1996), 171–85.

7   Jeremy Wilson, *Talk and Log: Wilderness Politics in British Columbia* (Vancouver: UBC Press, 1998), 144–45; Darren

NANCY JANOVICEK

R. Bardati, "Participation, Information, and Forest Conflict in the Slocan Valley of British Columbia," in Adkin, *Environmental Conflict and Democracy*, 103–22.

8   *SVCFMP Final Report*, 2–42.

9   Myrna Kostash, *Long Way from Home: The Story of the Sixties Generation in Canada* (Toronto: James Lorimer, 1980), 118; Katherine Gordon, *The Slocan: Portrait of a Valley* (Winlaw, BC: Sono Nis, 2004); *SVCFMP Final Report*, 2–22.

10   *SVCFMP Final Report*, 2–22.

11   This is based on my interviews; other historians working on the back-to-the-land movement note similar co-operation with longer-settled neighbours. Sharon Weaver, "First Encounters: 1970s Back-to-the-Land, Cape Breton, NS and Denman, Hornby and Lasqueti Islands, BC," *Oral History Forum d'histoire orale* 30 (2010): 1–30; Jinny A. Turman-Deal, "'We Were an Oddity': A Look at the Back-to-the-Land Movement in Appalachia," *West Virginia History* 4, no. 1 (2010): 1–32. See also the interviews conducted by Ryan O'Connor on Prince Edward Island: Alan MacEachern and Ryan O'Connor, *Back to the Island: The Back-to-the-Land Movement on PEI*, on NiCHE website, 2009, accessed 4 September 2015, http://niche-canada.org/member-projects/backto-theisland/home.html.

12   Matt Caver's chapter in this volume examines the origins of federal programs that targeted young people in another BC locality as well as criticisms of these programs.

13   "Signed by 30 [unnamed] residents," letter to the editor, *Nelson Daily News*, 16 June 1973.

14   Peggy Pawelko, "Death Threats, Hatred Cloud Slocan Valley," *Nelson Daily News*, 26 June 1973.

15   "The Sutherland Trials," *The Arrow*, August 1973; Michael Pratt, interview with the author. Nelson, BC, 21 October 2010. On the debates about the Free School, see Nancy Janovicek, "'The Community School Literally Takes Place in the Community': Alternative Education in the Back-to-the-Land Movement in the West Kootenays, 1959–1980," *Historical Studies in Education* 24, no. 1 (2012): 150–69.

16   Corky Evans, interview with the author, Winlaw, BC, 9 October 2010. Evans was born in Berkeley, CA, and moved to Vancouver with his wife and two children in 1969 or 1970. He worked as a longshoreman and logger and moved to the Slocan Valley to take a job as a surveyor in Castlegar in 1972.

17   These federal programs generally imposed an age limit of twenty-eight to qualify for the programs. Thanks to Kevin Brushett for this information.

18   *SVCFMP Final Report*, 2–42.

19   Josephine Reckart, *Public Funds, Private Provision: The*

Role of the Voluntary Sector (Vancouver: UBC Press, 1993).

20 Christopher Garrish, "Unscrambling the Omelette: Understanding British Columbia's Agricultural Land Reserve," *BC Studies* 136 (Winter 2002): 25–55.

21 Wilson, *Talk and Log*, 112–48.

22 Evans, interview.

23 *SVCFMP Final Report*, 3–91 (emphasis in original).

24 Ibid.; Evans, interview.

25 *SVCFMP Final Report*, 2–17.

26 Ibid., 2–26. Good agricultural land in the Arrow Lakes was also sacrificed to make way for hydroelectrical damming. See Joy Parr, *Sensing Changes: Technologies, Environments, and the Everyday, 1953–2003* (Vancouver: UBC Press, 2010), chap. 5.

27 The project commissioned these images to make sure that the report was accessible to all people in the valley. At the time, 37 percent of residents in the Slocan Valley were illiterate and would have struggled with the technical aspects of the report (Evans, interview).

28 *SVCFMP Final Report*, 2–29.

29 Ibid., 3–26.

30 Ibid., 3–47.

31 Ray Gill, *A Proposal for the Creation of a Public Working Circle in the Slocan Forest* (Nelson, BC: Nelson Forest District, 1955).

32 *SVCFMP Final Report*, 4–1.

33 Dave Richardson, "Local Resource Management," *Nelson Daily News*, January 1975.

34 *SVCFMP Final Report*, 1–2.

35 Ibid., 4–48.

36 Ibid., 4–14. Tina Loo emphasizes the continued importance of local woodsmen and conservationists despite the marginalization of local customs in government policy: *States of Nature: Conserving Canada's Wildlife in the Twentieth Century* (Vancouver: UBC Press, 2006).

37 In his examination of the interaction of humans and bears in national parks, George Colpitts argues that new knowledge of bear behavioural science changed perceptions about the relationship between bears and humans in natural spaces and led to a growing respect for "bear country": "Films, Tourists, and Bears in National Parks: Managing Park Use and the Problematic 'Highway Bum' Bear in the 1970s," in *A Century of Parks Canada, 1911–2011*, ed. Claire Elizabeth Campbell (Calgary: University of Calgary Press, 2011): 153–73. On changes in government policy, see Loo, *States of Nature*.

38 *SVCFMP Final Report*, ii.

39 Ibid., A-2.

40 Ibid., A-15.

41 Evans, interview; Bob Ploss, discussion with the author, Vancouver, 12 August 2010.

42 Williams, quoted in Wilson, *Talk and Log*, 145.

NANCY JANOVICEK

43 Evans, interview.

44 *SVCFMP Final Report*, i.

45 "Tripac Plans," November 1974, Slocan Valley Clippings file, Sean Lamb Archives, Nelson, BC; Les Storey, "Access to Timber a Rural Direction: Farming May Provide Stable Economy," *Nelson Daily News*, 4 March 1976; Phil Matheson, "Critique Given by Forester," *Nelson Daily News*, 27 November 1974; "Forestry Plan Made by Tri-Pac," *Nelson Daily News*, 26 November 1974; "Slocan Residents Named to Committee," *Nelson Daily News*, 12 May 1975.

46 "Logging in Your Back Yard: Where Have All the Forests Gone? (A Long Time Passing)," *The Arrow*, October 1974 (emphasis in original).

47 Braunstein and Doyle, *Imagine Nation*; Stuart Henderson, "Off of the Streets and into the Fortress: Experiments in Hip Separatism at Toronto's Rochdale College," *Canadian Historical Review* 92, no. 1 (2011): 107–33; Stuart Henderson, *Making the Scene: Yorkville and Hip Toronto in the 1960s* (Toronto: University of Toronto Press, 2011).

48 *SVCFMP Final Report*, v.

49 Ibid.; Wilson, *Talk and Log*, 266–70.

50 Bardati, "Participation, Information, and Forest Conflict," 120–22.

51 Wilson, *Talk and Log*, 266.

52 Integrated Land Management Bureau, "Kootenay-Boundary Land Resource Management Plan Implementation Strategy—Agriculture Fact Sheet," accessed 23 January 2013, http://archive.ilmb.gov.bc.ca/slrp/lrmp/cranbrook/kootenay/news/files/implementation_strat/agriculture.html#V.

53 HPCC mission statement, on Harrop-Procter Forest Products website, accessed 4 September 2015, http://www.hpcommunityforest.org/hpcc/. For discussions of the twenty-three-year campaign to found the co-op, see Colleen Shepherd and BCICS Research Group, "Harrop-Procter Community Co-operative" (Victoria: British Columbia Institute for Co-operative Studies, 2001), accessed 4 September 2015, http://www.uvic.ca/research/centres/cccbe/assets/docs/galleria/Harrop-ProcterCommunityCooperative.pdf; Harrop-Procter Watershed Protection Society, "Community Forest Pilot Agreement Proposal: Application for a Community Forest Licence," January 1999, http://www.hpcommunityforest.org/wp-content/uploads/2013/12/pilot_agree.pdf, accessed 23 January 2013, http://www.hpcommunityforest.org/home.html.

American Immigration, the Canadian Counterculture, and the Prefigurative Environmental Politics of the West Kootenay Region, 1969–1989[1]

Kathleen Rodgers

The "Genelle Three" were arrested and charged with obstructing a highway in late summer of 1978, after about forty people blocked a roadway leading to the site of uranium exploration in the hills behind the tiny working-class community of Genelle, just outside of Castlegar, British Columbia. In 1978, in light of rising uranium prices, industry advocates hailed the West Kootenay region, with its rich deposits, as the new uranium mining centre of BC. At the same time, the worldwide movement against nuclear armament also plagued the sector, inciting debate over a provincial moratorium on exploration. Thus, despite the small number of arrests and the remote location of the protests, the events garnered extensive media attention and crystallized the widely held view that uranium mining had no place in BC. In the spring of 1980, following the conviction of the protesters, the provincial government placed a seven-year moratorium on exploration and committed to ensuring that all uranium deposits in the province would remain undeveloped.

The widespread public support for the protesters and their political victory can be partially explained by the fact that the Genelle protests were seen by the local community and law enforcement alike as a truly grassroots resistance to mining. The legitimacy of the anti-uranium claims came in part from the respectability that the identities of the protesters demanded. As the provincial judge stated in his decision to convict the "Three"—but at the same time give them an absolute discharge—"I was particularly impressed with the credibility and integrity of all three accused. All three are working family men and upstanding members of the community. . . . They were motivated by the honestly-held belief that the exploration activities could endanger the health of their families and the community at large."[2] There was no question in these statements about whether the protests were the work of outside "agitators"; these were simply working-class people who cared about their families, their water, and their community.

That working-class residents in this small, remote community fought and won against industry appears to be an early victory for "environmental justice" advocates.[3] But the depiction of the events as a success of the marginalized working class fails to account for the circumstances that led to the mobilization of local residents. In reality, like so many of the environmental initiatives discussed in this volume, the Genelle protests took place against the backdrop of a burgeoning local counterculture. In the thirty years that followed the events in Genelle, West Kootenay life was punctuated by episodes of environmental contention—most notably by protests against logging—but also against mining and pesticides and in favour of wilderness preservation.

While a "vibrant counterculture" in the 1960s hinterlands of British Columbia might have seemed unlikely, its existence in a relatively isolated location arose from the migration of thousands of Vietnam War–era Americans to the West Kootenays and the political traditions they represented. Owing to these politics, the West Kootenays became home to a counterculture that embodied an

KATHLEEN RODGERS

overlapping set of values with respect to communalism, feminism, artistic expression, pacifism, democracy, a rejection of modern urban life, and a desire to go back to the land. Importantly for this chapter, members of the counterculture espoused an environmental critique of industry, represented in their politics and their own personalized quests for sustainable lifestyles.

The idea that environmentalism in Canada may have American roots is an unpopular sentiment in many academic circles[4]—and for good reason, as by the early 1970s grassroots environmentalism was not only a global movement but also well established in Canada. However, an embrace of environmentalism in rural regions remained exceptional even in 1978.[5] Understanding the nature of West Kootenay environmentalism, then, requires an understanding of the importance of this local counterculture, how it took shape, and how its environmental critique sharpened, becoming tailored to local issues and developing into organized environmentalism.

This chapter discusses the American origins of these local efforts but also demonstrates how the most successful campaigns of the West Kootenay counterculture were those that transcended these origins and fostered a broader community response. The politicization of collective goods such as water and old-growth forests provided a common focal point for community members and mobilized a broader public.[6] Still, the countercultural community—specifically, its leadership, expertise, ideas, and strategies—remained the epicentre of the resistance. For sociologist Wini Breines, the different ways of thinking and organizing within the New Left movements of the 1960s represented a form of "prefigurative politics," a rejection of conventional forms of political action. This case study of two episodes of environmental contention traces the ways in which the prefigurative politics of the American migrants were central to this counterculture and transformed social life in the West Kootenays.[7]

THE PREFIGURATIVE ENVIRONMENTAL POLITICS OF THE AMERICAN COUNTERCULTURE

American military conscription for the Vietnam War, combined with Canada's 1969 legislation allowing eligible immigrants legal admission to Canada regardless of their military status, drew more than 100,000 American men and women of draft age from the United States to Canada. For those who opposed the military draft, Canada was an obvious destination. But the war resisters were only one part of a much broader exodus of young people looking for alternative lifestyles; this retreat from militarism coincided with the hundreds of thousands of youth who joined communes or made the decision to go back to the land. With its vast stretches of inexpensive and virtually uninhabited terrain, British Columbia in particular provided a perfect context for Americans inspired by these ideals. That the young migrants would have a lasting impact in Canada is not surprising. Between 1967 and 1975, at least 19,000 Americans immigrated north to Canada each year, representing the largest number of American migrants to Canada since the United Empire Loyalists, and this rate has not since been exceeded.[8]

A distinct counterculture existed in Canada by the time the American migrants began to arrive. A counterpoint to the better-known American experiments, Canadian youth politics represented a similar emancipatory impulse. At the same time, the migration to Canada meant that the radical voices of New Left politics in the United States became loud and influential in Canada, contributing new political content to Canada's own prefigurative traditions. Frank Zelko makes this clear in *Make it a Green Peace!*, noting that even Greenpeace, Canada's greatest offering to environmentalism, was linked closely with American activism: "the organization may have started life in Canada but, to a large extent, its activist roots lie south of the 49th parallel."[9] Therefore, the American background of the migrants *was* important, and not merely because these Americans added critical mass to existing activism in Canada. As Jeff Lustig notes,

"Discontented youth ... agreed with Henry Miller that the American Dream had become an air-conditioned nightmare [and] were regularly told there were no alternatives."[10]

The implosion of New Left politics in the US in the late 1960s led to a greater quest for such alternatives and therefore to the rise of the personalized forms of politics of the 1970s, such as back-to-the-land and communal living. These were not simply extensions of the 1960s political movements, but rather, new expressions that "served as a transition to a new environmental politics in which the question of Nature could no longer be separated from the question of society itself."[11]

When these personalized politics combined with the circumstances of the migrants' lives in the context of the late 1960s, pockets of American counterculturalists took root in some of the most inhabitable but least populated rural areas around British Columbia, including the Gulf Islands, Bella Coola, Smithers, and the West Kootenays. Events such as those in Genelle demonstrate that in the ensuing years, at least in the West Kootenays, many of the young Americans took part in organizing local environmental campaigns and became active leaders, infusing local issues with environmental politics.

CONTEXTS OF COUNTERCULTURAL IMMIGRATION INTO THE WEST KOOTENAYS

The arrival and subsequent settlement of the counterculture in the region resulted from a particular constellation of economic and social factors. When the first few Americans began to trickle quietly into the West Kootenays in the late 1960s, the region was ripe for any form of development. This is not to say that the region was uninhabited. Indeed, the history of settlement in the region is rich and complex, consisting of multiple waves of immigration and economic development, as well as an Indigenous population that straddled the border with the United States. The largest wave of immigration accompanied mineral exploration in the late nineteenth century. European immigrants, many of them British, settled under the provisions of Canada's

Dominion Lands Act, establishing fruit orchards and farming the region since the early twentieth century. Perhaps most notably, there was a significant population of Russian-speaking religious and political refugees, the Doukhobors.

As was the case in many BC communities, settlement slowed following World War II as agriculture and mining declined; as a result, limited industrial development occurred and the population stagnated. When the federal and provincial governments promoted regional resource expansion following the war, hinterland regions like the West Kootenays embraced forest-based activities such as logging and pulp production, and some population growth occurred. But in 1965 just before Americans began arriving in great numbers, the population of the entire region, encompassing all municipalities and rural areas (known as the Regional District of Central Kootenay), was just forty-five thousand.[12]

The limited local economy had kept land prices low, creating opportunities for young people with few financial resources to settle and build lives on moderately arable land in a spectacular natural landscape. These same conditions had, in previous decades, attracted other migrants looking to establish intentional communities. The Argenta Quakers and the Doukhobors had both settled in the West Kootenays owing to the availability of land and the geographic isolation. These two groups were very distinct from each other—and from the American exiles—but shared important ideological and political beliefs. In both cases, the immigrants had fled their homeland because of their political and religious convictions.[13] Both communities shared values with respect to pacifism and agriculture. Based on these common worldviews and their own experience of exile, the Argenta Quakers and some members of the Doukhobor community provided practical and community support to the earliest American draft resisters. As ideological allies and back-to-the-land pioneers, the Doukhobors and the Argenta Quakers helped to create a hospitable environment for the establishment of the counterculture.[14] As the flow of American immigrants and Canadian adherents to the counterculture expanded and pockets of countercultural communities began to

KATHLEEN RODGERS

establish themselves, the region became a haven in which their prefigurative politics flourished. In this context, interactions between new arrivals and the established dissident groups began to decline, but they were frequently reactivated in the years following, during local political and environmental challenges.

STRATEGY AND RESISTANCE IN WEST KOOTENAY ENVIRONMENTALISM

As both an early and a successful campaign, the Genelle protests illustrate how the influence of the counterculture allowed the community of Genelle to leverage its very minimal economic and political power. Clearly, the provincial, national, and global anti-nuclear discourses assisted in the success of the Genelle protests.[15] The deaths of uranium miners in Elliot Lake, Ontario, in 1977, for instance, had provoked international condemnation of the industry, and in Vancouver, growing concern about nuclear contamination had inspired the mayor to declare a "Trident concern week" as environmental groups in both the US and Canada actively protested the Trident nuclear submarine base in Washington state.[16] For some in the Kootenays, a prospective uranium mine meant economic development in a perpetually depressed region, and as such, local officials supported the project. In a speech at the local college, for example, the regional representative of the federal department of mines stated that uranium mining was much more difficult in other geographic locations. "We are lucky," he commented, "we live in a uranium province."[17] Thus, while the broader political context appeared to favour anti-nuclear protest, the local economic and political context was less propitious.

Local interest in the issue was sparked when Vancouver consultants for a Toronto-based mining company began taking samples from the hills behind Genelle, in the China Creek watershed, in the fall of 1977. Shortly after blasting began, and months before protest barricades were erected to prevent the engineers from exploring the territory, the Kootenay Nuclear Study Group (KNSG) formed in response to the exploration. Members of the KNSG were not from

Genelle. Most were former Americans who lived in the countercultural stronghold thirty-five kilometres away in the Slocan Valley. Several also had previous experience in the political movements of the 1960s US protest wave. Members of the group had heard rumours about the exploration in Genelle and had concerns about mining exploration in their own watersheds. In the subsequent months, members of the KNSG informed themselves of the evidence and arguments put forward by anti-nuclear advocates. The group launched a campaign to question the activities of government and corporate mining interests in the region—meeting with Jim Chabot, the provincial minister of mines, and the regional mines inspector—and to amass evidence in support of their belief that uranium mining was hazardous to the water supply and the health of the local population. The group wrote letters of protest, documented the exploration with photographs, invited experts to speak on the topic, and in the spring of 1978 began to liaise with members of the community in Genelle.[18]

The coordinator of the KNSG, a young American named Jim Terrall, spoke at a Genelle town meeting in order to explain "the dangers of radiation pollution in the drinking water and from a possible future mining operation." Indicating the extent to which mobilization around environmental issues was not a common feature of local life, one account of the meeting noted that "up till then the people had been more concerned about dirt in their drinking water; radiation was a new concept to them."[19] Aside from their involvement in local union politics, the people of Genelle had little experience in civic action. However, the fact that explorations were sponsored by a Toronto-based consortium and a Vancouver-based engineering company was not lost on the assembly, and those in attendance resolved to form the Genelle Concerned Citizens Action Committee (GCCAC). The spokesperson and de facto leader of the group, Tom Mackenzie, an active union organizer, lent the group credibility among the local population.

In subsequent months, the GCCAC and the KNSG worked together closely, meeting with officials and planning a barricade to prevent mining equipment from passing and drilling inside the watershed.

When it became clear that the mining consultants would proceed with drilling, the KNSG and local residents began to talk of protest. Members of the KNSG were committed to the idea that social change should be achieved through nonviolent means. M. L. Burke, a member of the KNSG, recalls that within the group "there was a definite consensus that we should be doing this through nonviolent means."[20] As summer arrived and a barricade was constructed, the group invited members of the Pacific Life Community (PLC) of Vancouver, a California-based peace organization dedicated to the use of nonviolence in the pursuit of nuclear disarmament.[21] The PLC had become well known in Vancouver and Washington State for its anti-nuclear stance on the Trident nuclear-missile base in Bangor, WA. Most notably, PLC had organized acts of mass civil disobedience against the Trident base, orchestrating the arrest of thousands of protesters who scaled the facility's fence.[22] Well versed in the philosophy of nonviolence, the PLC presented the first of a number of workshops on the principles and practice of nonviolent resistance and civil disobedience to the residents of Genelle. The workshop advocated classic principles of nonviolent resistance:

> exercises were given on "listening" and on defining and communicating one's concerns and objectives. There was role-playing practice for a number of confrontation situations with people acting the parts of "protestors," [sic] "police" and "drill-crew members." . . . Exercises in quick consensus decision-making were given and the instructors cautioned that "violence" and "non-violence" can <u>never</u> be mixed with any hope of success—use violence at any time, they said, and you destroy all your credibility and lose any sympathy you may have gained.[23]

About thirty people attended the workshop: members of the KNSG and residents from other locations in the Slocan Valley. Despite the fact that no Genelle residents were involved at this stage, members of the KNSG went to the barricade and conducted their own workshop on techniques of nonviolent resistance. Later, when a confrontation between protesters and mining representatives appeared imminent, the demonstrators employed these principles: representatives of the group informed police that they did not intend to engage in any violence or to obstruct the duties of the police. However, they also conveyed that "if improving the law might have to involve breaking it, well, there was a long and honourable tradition for this . . . and the people of Genelle were individually examining their hearts and their consciences."[24]

In the early summer, techniques of nonviolent resistance played a central role in the eventual outcome. On the morning of the arrests of the Genelle Three, the assembled protesters selected those who would be arrested. Strategically, they decided that only residents of Genelle should be detained. Having identified a narrow spot in the access road, the protesters commenced with their sit-in, forming a human chain to prevent mining equipment from passing into the hills where drilling was set to occur. Thirty-five kilometres away at the regional headquarters of the department of mines a delegation, including the coordinator of the KNSG, threatened their own sit-in when the water rights inspector refused to see them. When the meeting eventually took place, the official lectured them, pointing out that the uranium engineers had the right to conduct their explorations and that any further action would lead to arrests. When the group reconvened at the barricade and the bulldozer attempted to proceed, the group again formed a human chain. The police moved in and reluctantly arrested Herb McGregor, Eric Taylor, and Brent Lee, the three nominated Genelle residents, for obstruction of a public roadway.

The strategy employed by the organizers was a clear success. The fact that the people of Genelle were willing to pay the price of jail time to protect their water drew support throughout the province. In Vancouver, on the day following the arrests, the Society Promoting Environmental Conservation (SPEC) and the Canadian Coalition for Nuclear Responsibility held a joint demonstration in front of the department of mines office to show support for the people of Genelle. Because of the arrests, the barricades swelled with protesters throughout the summer as the trial of the Genelle Three proceeded. In late January 1979, on the day following the summary legal arguments for the Three, the province bowed to both public pressure and the momentum of the Genelle protests and announced its intention to hold a Royal Commission of Inquiry into Uranium Mining (RCIUM).[25] In his decision to convict the Three while handing down an absolute discharge, presiding Judge Bruce Josephson reflected on his respect for the individuals involved and on their legitimate use of civil disobedience. Drawing on comments from the then chief justice of the Manitoba Court of Appeal, Josephson noted, "a society places a high value on dissent and other peaceful challenges to the rule of law."[26] Shortly after the conviction of the Three, members of the RCIUM visited the site of exploration in Genelle. In light of the official scrutiny, the growing hostility toward uranium exploration in the province, and the "favourable" discharge of the Three, the consultants called off their exploration and commented, "We're not in the business of fighting people."[27]

Genelle, with its population of just five hundred people in 1978, had successfully used the traditions of nonviolent civil disobedience to defend the community's water from a powerful representative of industry. But while the strategic deployment of civil disobedience points to the influence of standard countercultural strategies, the use of such tactics also highlights the cultural ferment taking place. The events in Genelle represented the coming together of members of the counterculture with the region's longer-term residents, a merger that was not always comfortable. Given the large influx of young counterculturalists—and the fact that they were American, in particular—conflict over values in the region was long-standing and in fact had increased cohesion within the counterculture. For the counterculturalists, the use of civil disobedience was a valiant, time-honoured tradition and a legitimate expression of discontent. From the perspective of those without roots in this tradition, civil disobedience still amounted to breaking the law. As evidence of this, one of the Genelle Three wrote a letter to the editor of the local paper following his conviction. While apologizing to the RCMP, the letter writer conveys his personal struggle with the use of civil disobedience:

> To my good friends the RCMP, sorry for any inconvenience. You did your job and you did it well, but try to realize when there's 50 of you and a little pregnant mother stands up to your chief to tell him she will lie down anytime in front of a car, there just has to be a reason. Laws were made by man to serve the majority. When they get old and no longer do this, but rather licence a few to jeopardize a whole village, it is time they were changed.[28]

The Genelle protests were not the first episode of civil disobedience initiated by the counterculturalists, but they were the first in which members of the countercultural community and established residents came together. Despite never taking centre stage in the Genelle conflict, the counterculture brought forward ideology, tactics, leadership, and a cohesive community of people motivated and willing to promote local environmental concerns. The counterculturalists also supported the campaign financially. For instance, community organizers held a fundraising event to help pay the legal fees of the Three. The event included auctioning a homemade cake—a replica of the Three Mile Island nuclear power plant.[29] The cake was donated by Sally Lamare, just one of the American expatriates who had gone back to the land in the region. Unlike residents in other communities, where locals did not possess a tactical repertoire allowing them to successfully leverage their minimal power, those in Genelle employed the toolkit and resources of the resident counterculture. For this reason, the trajectory of conflicts over environmental rights in the West Kootenays is different than that of many other resource-based communities in British Columbia.[30]

KATHLEEN RODGERS

ORGANIZING FOR A HERBICIDE/PESTICIDE-FREE COMMUNITY

West Kootenay environmentalism is most renowned for its contentious anti-logging protests of the 1990s. However, some of the earliest and most successful environmentalist protest efforts involved campaigns to end herbicide and pesticide use. As with the Genelle protest, the leadership, organizations, and tactical repertoire of the anti-herbicide/pesticide campaigns were firmly rooted in the local counterculture. The broader success of the West Kootenay anti-herbicide/pesticide activism arose from the fact that the issue involved a broader public good. By framing herbicide and pesticide use as an assault by industry on local watersheds and local decision making, the protests spread to a much larger segment of the population. The trajectory of the anti-herbicide/pesticide activism of the 1980s had first taken shape much earlier, as a fervent environmental consciousness infused the small back-to-the-land communities. The prominence of the forest industry in the region had drawn attention to the impact of forestry practices and other industrial behaviours on the quality of local water. In turn, growing awareness of these trends facilitated the growth of organizations that later served as the launching pad for subsequent environmental activism.

By the time the first counterculturalists arrived in the West Kootenays, it was a well-established centre for highly industrialized logging activity, with a small number of companies controlling rights to timber extraction and production. Logging loomed large in the economy and politics of the region. Whether it was through their employment in the new tree-planting industry or as loggers, members of the local counterculture quickly recognized the impact of forest practices on the aesthetics of the local landscape and the quality of their water.[31] With a fifty-thousand-dollar federal Local Employment Assistance Program (LEAP) grant, a local committee spent two years developing the Slocan Valley Forest Management Project (SVFMP); in 1975, it released a report evaluating standard forest practices and outlining a sustainable approach to local forestry. The committee's final report garnered extensive local support from a wide range of stakeholders, but efforts to implement the plan ultimately failed.[32] The process had nonetheless identified and articulated the community's collective environmental interests while also leading to new divisions within the community about how best to achieve their goals. While some favoured a continued institutional approach, a desire for direct action also began to take shape.

Out of the ashes of the failed forestry reconfiguration process emerged a number of activist-oriented groups with specific mandates to protect water and wilderness. First, in response to ongoing concerns about water quality, watershed protection groups formed throughout the West Kootenays. The first of these, the Perry Ridge Water Users Association (PRWUA), arose in 1981 in the Slocan Valley, where the densest and most active countercultural population resided. The PRWUA was among the first watershed associations in the province. Shortly after, the Slocan Valley Watershed Alliance (SVWA) formed, quickly becoming a powerful environmental and political advocate in the region and beyond. The SVFMP gave new life to the idea of creating a land conservancy in the Slocan Valley; the Valhalla Land Conservancy, later the Valhalla Wilderness Society (VWS), the brainchild of three young Americans (Ave Eweson, Grant Copeland, and Richard Caniell), came together for this task in 1975.[33] After extensive lobbying of the provincial government, the VWS ensured the creation of Valhalla Provincial Park, and it continued to build a strong membership base and provide leadership in the BC wilderness protection movement. Thus, while environmental consciousness had been growing in the region well before the founding of the watershed societies and the VWS, these groups became the organizational basis for environmental consciousness and protest mobilization.[34]

The groups monitored local forestry practice and engaged with industry officials on their use of pesticides/herbicides in the region. In the early 1980s, the residents of the Slocan Valley and nearby Argenta became aware of the intention of the BC Ministry of Forests to use products such as Roundup (glyphosate) to reduce excess brush and of BC Hydro's routine use of the herbicide to clear areas below power lines. In Argenta, residents formed the Nonviolent Action Group (NAG) to engage in direct action campaigns against pesticide/herbicide use in the region. In this context, beginning in 1985, the VWS, the SVWA, and the NAG launched appeals of Ministry of Environment pesticide/herbicide permits with the BC Provincial Appeal Board.[35] Without exception, board members ruled that the Ministry of Forests, BC Hydro, and CP Rail had demonstrated that their use of herbicides posed no threat to "man or the environment," and the appeals were unsuccessful. In the earliest appeal, the panel concluded that, "notwithstanding the views to the contrary expressed by a number of sincere, dedicated local environmentalists, the treatments authorized under the Permits are justified and will not cause any unreasonable adverse effects."[36] The organizations, buoyed by the belief that the local communities remained concerned about the possibility of adverse effects, lobbied the local representatives of the Regional District of Central Kootenay (RDCK) to establish the West Kootenay region as an herbicide/pesticide-free zone.[37] The RDCK, a strong supporter of local decision making, decreed that

> the use of all pesticides/herbicides by the Ministry of Forests, the Ministry of Highways, BC Hydro and Power Authority and West Kootenay Power and Light Company be immediately discontinued and the boundaries of electoral areas A, B, D, G, H, I and J be recognized as pesticide/herbicide-free zones.[38]

The RDCK's proclamation did not prevent the environment ministry from permitting industrial spraying in the region, but it did serve as a platform for mobilization, setting the stage for three subsequent years of direct action against pesticide/herbicide use in the region. To better coordinate the direct action elements of these campaigns, an offshoot of the SVWA formed: Kootenay Citizens for Alternatives to Pesticides (KCAP). With continuing failures at the provincial appeals board, dissent grew within the countercultural communities, and in the summer of 1986, members of NAG, frustrated by the continued awarding of permits, declared, "people are opposed to the use of pesticides by public agencies on public land. . . . The Regional District of Central Kootenay has declared this area a pesticide free zone. . . . Public servants must respect the will of the public."[39] Waving banners reading "pesticide-free zone," protesters in the Slocan Valley placed their vehicles across roadways to prevent CP Rail from spraying Tordon 101, and the NAG blocked the road by which Ministry of Forests vehicles could access their herbicide warehouse and surrounded helicopters loaded with Roundup to prevent them from taking flight.

In 1987, following the news that CP Rail's permit to spray Spike 80W (tebuthiuron) on the railways of the region would stand, activists mobilized a much broader campaign. The now established network of anti-pesticide, watershed, and environmental organizations in the region coordinated a multifaceted campaign to involve the largest possible subsection of the Slocan Valley population. At this time there had been no successful challenge to the use of pesticides by Canada's railway corporations, but the campaign drew momentum from revelations in Sault St. Marie, Ontario, that CP Rail would pay millions of dollars in cleanup and restitution after Spike had seeped from the rail bed into private yards, killing lawns and trees and seeping into basements.[40] The opposition in the Slocan Valley was fierce and effective, launched by SVWA and KCAP but drawing on the support of the RDCK, local schools, unions, and countercultural institutions. The groups encouraged citizens to join the campaign by signing petitions and writing letters to the minister of the environment and to the regional pesticide control manager (and hundreds of letters were indeed written). But members of the SVWA maintained their commitment to the idea that if these legal channels did not work, "illegal and possibly violent actions would be likely," commenting that "they'll have to put us in jail to get us out of the way."[41] As in earlier campaigns, leadership and ideas from within the counterculture were central. In one letter, written by Vietnam War veteran Philip Pedini and addressed to the local pesticide control manager, Stuart Craig, Pedini used his experience to frame his opposition to CP Rail's use of Spike:

KATHLEEN RODGERS

Around the base at Bien, Hoa Vietnam, from horizon to horizon, the land was "defoliated" from herbicide sprays. . . . Vietnamese women had so many stillborn babies, so many babies born with severe birth defects. . . . I cry for the Vietnamese people. I cry for my friend whose stillborn baby had no brain. . . . Who do you cry for Mr. Craig? Think of your friends and relatives. Your loved ones. . . . I'm asking you to explore the doubts you must have about pesticides. . . . The people of the central Kootenay live in fear of the spray truck contaminating our gardens, our favourite fishing and swimming holes, our livestock and our waters. We are afraid of what Spike might do to our children and ourselves.[42]

Pedini also instructed Craig to consult Rachel Carson's *Silent Spring*, the book widely viewed as the intellectual impetus for the modern environmental movement.

With CP Rail's permits to spray sections of the region's railway still in effect, the persuasiveness of such arguments began to take hold. In early July 1987, the pesticide control manager toured the contested spray area, where he was met by seven hundred protesters. In the days following, he recommended the cancellation of the pesticide permit, on the grounds that the spraying was too close to the water.[43] This was the first victory for anti-herbicide activists in the region and one of only a handful in the province. Commenting that the permit process was "screwed up," a member of the local Environmental Appeal Board reflected on the victory of the protesters: "one of the things that makes me particularly pleased about working in this area is that people do question authority. They don't automatically accept the fact that just because the government made a decision that it was necessarily the right decision."[44] But the victory was incomplete. In neighbouring communities where much less resistance to the spraying had been demonstrated, the permits remained in effect and the spraying proceeded. In the following days, the editor of a local paper wrote, "the only conclusion that can be drawn from all this is that the protests worked. Castlegar residents didn't get out and make their concerns heard. Slocan Valley residents did. It's as simple as that."[45]

Encouraged by their own success, the activists of the KCAP, NAG, and SVWA mobilized the communities where the permits still stood. In the following twelve months, blockades were constructed in more communities around the region. In 1988, in Nelson where CP Rail intended to spray the tracks, the city council joined the regional district in applying for a court injunction to stop the railway from spraying Spike in and around the city.[46] When the efforts of officials appeared to be failing, local residents came forward and blocked the tracks.[47] Rather than risk the publicity of arrests, the CP Rail spray truck turned and left, and the company announced later in the day that it would abandon all efforts to spray the Nelson-Creston line. Following this victory, the director of the RDCK commented publicly, and ironically, on the failure of formal political channels to respond to the desire of people to keep pesticides out of their community and on the central role that the activism played in the successful outcome:

> I would like to apologize for the futile efforts that we made to help you. I apologize for the lack of support from our learned judges, who may know the law, but know less about environmental matters than the least informed of you here tonight. I apologize for the area MLA's [members of the provincial legislative assembly], our representatives, who helped us not at all. I apologize for the . . . civil servants whose great salary we pay and who forever side with the companies and manufacturers who would drench us with their poisons. . . . Yours is a very great victory. Your agonizing moments, sleepless nights, lost time, and above all, your concerned dedication to a good cause, has brought us all a victory. The spark that you have blown into a great fire, burns now across the province as others become aware of what people can do and what politicians and the law cannot.[48]

KATHLEEN RODGERS

After a three-year battle, by the spring of 1989, CP Rail had no active permits to spray its tracks in the West Kootenays. That fall, railway officials even took participants in the anti-Spike campaigns on a tour of the tracks in their newly developed steam machine designed to eliminate weeds using non-chemical technology.[49]

By the time the communities of the West Kootenays had come to-gether to resist herbicide/pesticide use, this wing of the environmen-tal movement was in full force throughout North America. SPEC had been raising awareness of the dangers of herbicide and pesticide use by BC industry since the early 1970s, and the Union of BC Indian Chiefs had sounded alarm bells about chemical use in Aboriginal commu-nities since the 1980s. Spurred by the success of the West Kootenays campaign, other rural communities launched appeals and engaged in direct action against the use of herbicides in their communities. Yet as recently as 2003, SPEC was fighting unsuccessfully against CP Rail's use of herbicides in the Vancouver region.

WEST KOOTENAY ENVIRONMENTALISM BE-YOND THE COUNTERCULTURE AND BEYOND AMERICAN IMMIGRATION

Beginning in the 1980s, community-level logging conflicts became a regular feature of life in rural British Columbian communities; it was a period that earned the moniker of the "War in the Woods." Political scientist Jeremy Wilson comments that these forest con-flicts transformed politics in a province that remains an otherwise "frustratingly inert democracy."[50] These dynamics were no less pro-nounced in the West Kootenays, where organized environmentalists engaged in a decade of logging protest. However, by the 1990s, pro-vincial environmental politics were shaping the dynamics of local en-vironmental struggles with public relations teams hired to crush the public image of environmentalists, successfully pitting labour against environmentalism. The population of the West Kootenay region had also diversified, and many of the activists on the frontlines of the barricades and behind the scenes of environmental advocacy were not the countercultural pioneers of the sixties and seventies. Still, the influence of this generation endures, through the established organizations, the shared history and traditions of dissent in the region, and local commitments to sustainable environmental lifestyles.

A criticism regularly levelled at the Americans who came to Canada in the 1960s and 1970s is that they were merely looking to "drop out." The story of the counterculture in the West Kootenays exemplifies the oversimplification of such narratives. Owing to the particular economic and social conditions presented by the West Kootenay region, the seemingly fanciful ambition of creating a non-violent, sustainable, democratic community seemed possible to the migrants as the population of like-minded newcomers reached a critical mass. The prefigurative impetus of the migrants to produce social change through personal and collective endeavours meant that community members formed enduring institutions and voluntary organizations, and launched repeated and successful environmental campaigns. Local environmental conflict reveals the importance of these origins; today, many leaders in the countercultural community remain active in local politics and at the helm of organizations, and they act out their commitment to a range of countercultural values through the politicization of their daily lives.

NOTES

1   Some of this material appeared in a different form in my book, *Welcome to Resisterville: American Dissidents in British Columbia* (Vancouver: UBC Press, 2014). I thank the press for their permission to publish it in this volume.

2   "Sentencing Transcript," *Nelson Daily News* (hereafter *NDN*), 12 April 1979.

3   See Robert D. Bullard, *Dumping in Dixie: Race, Class, and* *Environmental Quality* (Boulder, CO: Westview, 1990).

4   This concern stems largely from the very recent emergence of environmental history as a field in Canada rather than an emphasis on Canadian exceptionalism. See, for example, George Warecki's excellent book, *Protecting Ontario's Wilderness: A History of Changing Ideas and Preservation Politics, 1927–1973* (New York: Peter Lang, 2000). See also

Alan MacEachern, "Voices Crying in the Wilderness: Recent Works in Canadian Environmental History," *Acadiensis* 31, no. 2 (2002): 215–26.

5   Maureen Reed, *Taking Stands: Gender and the Sustainability of Rural Communities* (Vancouver: UBC Press, 2003); Tina Loo, *States of Nature: Conserving Canada's Wildlife in the Twentieth Century* (Vancouver: UBC Press, 2011), 191–92.

6   The counterculture also inspired significant conflict over competing visions of life and economy in the region. Nancy Janovicek explores this theme in this volume, as I do in *Welcome to Resisterville.*

7   Wini Breines, *Community and Organization in the New Left, 1962–1968: The Great Refusal* (New Brunswick, NJ: Rutgers University Press, 1989).

8   John Hagan, *Northern Passage: American Vietnam War Resisters in Canada* (Cambridge, MA: Harvard University Press, 2001).

9   Frank Zelko, *Make It a Green Peace!: The Rise of Countercultural Environmentalism* (New York: Oxford University Press, 2013), 10.

10 Jeff Lustig, "The Counterculture as Commons: The Ecology of Community in the Bay Area," in *West of Eden: Communes and Utopia in Northern California,* ed. Iain Boal, Janferie Stone, Michael Watts, and Cal Wislow (Oakland, CA: PM Press, 2012), 31.

11 Robert Gottlieb, *Forcing the Spring: The Transformation of the American Environmental Movement* (Washington, DC: Island Press, 2005), 148.

12 "Regional District and Municipal Census Populations, from 1941 to 1986," on BC Stats website, accessed September 21, 2015, http://www.bcstats.gov.bc.ca/StatisticsBySubject/Census.aspx.

13 Jean Barman, *The West beyond the West,* 2nd ed. (Toronto: University of Toronto Press, 2007), 153.

14 I explore the welcome provided by these groups in *Welcome to Resisterville.*

15 Residents of Clearwater, north of Kamloops, also called for an inquiry into uranium mining after the announcement of plans for a mine in the region; however, it was only in Genelle that citizens engaged in civil disobedience.

16 David Gersovitz, "Pollution's the Price of Prosperity," *Montreal Gazette,* 20 February 1978; "Vancouver Protests Sub Base," *Calgary Herald,* 19 November 1975.

17 Quoted in Joan Reynolds, "The Genelle Diary," unpublished manuscript, 1979, Selkirk College Library, Castlegar, BC, p. 14.

18 Kootenay Nuclear Study Group, *Newsletter* 1, no. 1 (1979).

19 Reynolds, "The Genelle Diary," 17.

20 Mary-Lynn Burke, interview with the author, August 2011.

21 See Barbara Epstein, *Political Protest and Cultural Revolution: Nonviolent Direct Action in the 1970s and 1980s* (Berkeley: University of California Press, 1991).

22 Ibid., 201.

23 Reynolds, "The Genelle Diary," 26.

24 Ibid., 28.

25 A means of dealing with the public outcry, the commission was announced a few hours before a public demonstration opposing uranium mining was scheduled for the steps of the BC Legislature. Public hearings to be held throughout the province were abruptly cancelled when then Premier Bill Bennett issued a seven-year moratorium on uranium mining. Despite the abrupt end to the hearings, the commissioners, with Dr. David V. Bates as chair, released their report in October 1979, in which uranium mining was identified as a threat to public drinking water. The moratorium was not renewed in 1987. David V. Bates, J. W. Murray, and V. Raudsepp, "The Commissioner's First Interim Report on Uranium Exploration," *Royal Commission of Inquiry: Health and Environmental Protection—Uranium Mining* (Vancouver, BC, 1979).

26 "Sentencing Transcript."

27 Reynolds, "The Genelle Diary," 143.

28 "Genelle Man Accepts Decision," *Castlegar News*, 8 April 1979.

29 Burke, interview. The Three Mile Island accident, which occurred in Pennsylvania in 1979, resulted in the release of radioactive waste into the Susquehanna River and the evacuation of local residents. To date it is the worst nuclear accident in North America.

30 However, note the similarities to the Denman Island experience, as covered by Sharon Weaver in this volume.

31 Valley Resource Society, *Slocan Valley Community Forest Management Project Final Report*, 2nd ed. (Winlaw, BC: Slocan Valley Resource Society, 1976).

32 See Nancy Janovicek's comments in this volume on the eventual adoption of this plan by the province.

33 Richard Caniell, interview with the author, 27 June 2009. For a detailed history of the VWS, see Jeremy Wilson, *Talk and Log: Wilderness Politics in British Columbia, 1965–1996* (Vancouver: UBC Press, 1998).

34 The SVWA was also active provincially, leading "For the Love of Our Waters" workshops and serving as a central organizer for anti-logging campaigns throughout the 1980s and 1990s.

35 Replacing the Pesticide Control Board, the British Columbia Environmental Appeal Board is an independent agency established in 1981 with a mandate to

hear appeals under the Pesticide Control Act, Waste Management Act, Water Act, and Wildlife Act. These four acts have been amended and replaced in the intervening years.

36 Environmental Appeal Board, "Judgement: Appeal against Pesticide Control Act," Ministry of the Environment, Appeal No. 85/10 PES, 1985, J-2, accessed 28 May 2015, http://www.eab.gov. bc.ca/1985ALLDECList.htm.

37 Incorporated in 1965, the Regional District of Central Kootenay is one of twenty-seven regional districts in British Columbia.

38 "Minutes," RDCK, 23 June 1985, Philip Pedini personal papers.

39 "Spraying Blocked by Lardeau Residents," NDN, 17 June 1986.

40 "Soo Cleanup Over, CP Rail Says," Globe and Mail, 24 December 1988.

41 Vancouver Sun, 25 May and 17 August 1988.

42 Philip Pedini to Stuart Craig, 17 June 1987, Philip Pedini personal papers.

43 "The Squeaky Wheel," Castlegar News, 12 July 1987.

44 "Permit Process Is Screwed Up," NDN, 28 July 1987.

45 "The Squeaky Wheel."

46 "Nelson Joins Spray Protest," Vancouver Sun, 26 May 1987.

47 "Herbicide Protest," Vancouver Sun, 18 June 1987.

48 "Spike Protesters Celebrate Victory," NDN, 1 September 1988.

49 Dave Polster (former scientist and environmental supervisor for CP Rail), personal communication, 30 March 2011.

50 Wilson, Talk and Log, xxix.

Countercultural Recycling in Toronto: The "Is Five Foundation" and the Origins of the Blue Box

Ryan O'Connor

Blue box recycling is big business in Ontario. In 2010 the program serviced 95 percent of the province's homes; in the process, over 900,000 tonnes of materials—or 68 percent of the province's total waste—was diverted from landfill. Managed by Waste Diversion Ontario, which was created by an act of the provincial Parliament, the program's costs are evenly distributed among the municipalities and Stewardship Ontario, a not-for-profit organization funded by the companies whose products are collected.[1] Long renowned as one of the world's most comprehensive and effective recycling programs, Ontario's blue box initiative was recognized in 1989 with an environmental award of merit by the United Nations.[2] Use of the blue box has not been confined to Ontario; it has been adopted in hundreds of municipalities throughout Canada as well as the United States, Europe, and Australia. The successful, and widespread, adoption of the blue box belies that object's rather humble origins. This chapter examines the story of the Is Five Foundation (IFF). Founded in 1974 according to countercultural principles, the IFF was a beehive of activity that undertook a plethora of initiatives. The group found its greatest success in the field of recycling.

During the 1960s and 1970s, Toronto was home to a rich countercultural community. Stuart Henderson's *Making the Scene* documents how the Yorkville neighbourhood was the premier Canadian hippie destination in the 1960s.[3] Grant Goodbrand has documented the sometimes strange story of the Therafields psychoanalytic experiment that developed in the Annex and grew to be the country's largest commune.[4] Yorkville would eventually be gentrified, replacing its hippie-oriented cafés and coffee houses with upscale shopping, while Therafields has long since sold off its once extensive properties. The IFF, then, provides something unique. On any given day throughout Toronto—or elsewhere in Ontario and around the world—people view a curbside reminder of this countercultural organization's legacy.

Exploring the experience of the IFF furthers our growing understanding of Toronto's countercultural past as well as its contributions to Canada's environmental history. This chapter will also shed light on the relationship between the counterculture and business. In one respect, the IFF operated a number of business ventures, aimed at generating the income necessary to continue operations and sufficient for the members to earn a living. In creating the blue box, the IFF and its spinoff organizations worked closely with a variety of corporations, most notably Laidlaw Waste Systems Ltd. While we tend to think of the counterculture as being, by nature, averse to "big business"—how many times have we heard critiques of former hippies who "sold out" and began working for "the man"?—the story of the IFF reveals this notion to be rather simplistic. Just as some of the key innovators in the personal computer industry had countercultural backgrounds, the IFF had similar entrepreneurial success.[5] The IFF demonstrates how a countercultural organization can, through natural developments and happenstance, develop into a mainstream corporate entity.

JACK MCGINNIS AND THE FOUNDING OF THE IS FIVE FOUNDATION

Born on January 3, 1947, and raised in the prosperous Cleveland suburb of Solon, Ohio, Jack McGinnis was the third of four children. From an early age, McGinnis showed the signs of a sharp intellect, which was verified when tests conducted in high school revealed he had a genius-level IQ. A mischievous youth—a characteristic his sister attributes to the boredom of life in a small town—he developed an interest in writing while in high school. Having worked at the school newspaper, during which time he won awards in various national writing competitions, he decided to study journalism at the University of South Florida in Tampa.[6]

In sharp contrast to his life in staid Solon, McGinnis found Tampa to be an intellectually nourishing environment. McGinnis enjoyed the cultural and ideological diversity found on campus and in the surrounding city. While he continued writing, he engaged in other activities such as experimental theatre. Described as a "free spirit" by some and a "hippie" by others, he, like so many of his contemporaries, underwent a dramatic physical transformation in the late 1960s. Whereas his hair had previously been cropped short and he had been inclined to wear khakis—the traditional preppy look—he began to grow his hair and beard long and to wear blue jeans.[7]

The cynical political climate of the late 1960s, marked by the escalating war in Vietnam and Richard Nixon's election as president, began to take its toll on McGinnis. A teenage bout with spinal meningitis left him draft exempt. His friends were not as fortunate, leading some to move to Canada as war resisters. Shortly after graduating in 1969, McGinnis and his first wife, Michelle, made the decision to follow his friends north, believing that it would be unconscionable to continue living in the United States. They initially lived in the Niagara region, where he worked as a photojournalist. In 1971, McGinnis moved to Toronto and began working a variety of odd jobs, which included driving a bookmobile and a taxicab. These work experiences

proved to be less than fulfilling and were the impetus for some inspired thinking. As he recalls,

> I had realized years before that I wasn't really cut out to
> work for anybody else. That wasn't my lot in life, not what
> I enjoyed. And so I started something in the early seven-
> ties, just a small business, and was really successful, in
> those terms anyway, but then had a huge shock which was
> the realization [that] as much as I didn't like working for
> somebody, I also didn't like the idea of somebody work-
> ing for me, which was more of a surprise than the first one
> was. So what came out of that was a strong desire to find a
> way to work with people, and I didn't really know what it
> exactly was at that point, I just knew it was looking for a
> way to take on something with other people in a teamwork
> relationship, not in a traditional business way. That was the
> stronger thing for me: I hadn't really set out to be an envi-
> ronmentalist or to be a recycler or anything else. I set out to
> be a "worker co-operative" person.[8]

The desire to create a worker co-operative resulted in the creation of
the "Is Five Foundation." The choice of a rather unusual name was
deliberate, as McGinnis felt it would create a natural opportunity
to explain the organization's purpose. The name was derived from
two sources: Buckminster Fuller's concept of synergy, and a book of
poetry, is 5, by E. E. Cummings. According to McGinnis, "The idea
was to find a way for people to work together so that it was exciting
and inspiring, and so ultimately the whole would be greater than the
sum of the parts, and what we did together would be more than if
we worked on our own."[9] In essence, the aim was to empower people
through co-operation. "We wanted to tell people there was a problem," explains McGinnis, "but the solution was them in their own
home and their own lifestyle. So it was very much people working
together within the group, and trying to find practical ways to ask

people in their own home and eventually in their workplace to do things differently."[10]

The IFF established itself as a non-profit, registered charity and began operation as a collective, with its seven initial members all participating in decision making. Its first effort was a roadside, multi-material pickup that operated weekly in Toronto's east-end Beaches district. Launched in January 1975, Project One Recycling filled a sizable void. Despite a growing awareness of the benefits of recycling, a by-product of the emerging environmental consciousness of the period, few options existed for those wishing to participate in this activity. Due to the high value of newsprint, the city of Toronto began experimenting with paper pickups in 1971.[11] Those wanting to recycle items such as metal, plastic, or non-newsprint forms of paper were forced to seek out depots where they could drop off their materials. The depots, which tended to be staffed by volunteers, were often short-lived operations, fluctuating with market prices for reclaimed materials and the ability to procure government grants to cover operating costs.[12] Prior to beginning operations, IFF members travelled door to door publicizing the program while McGinnis drove the organization's lone vehicle, a pickup truck. Project One Recycling focused on practical research. According to the IFF, "It is designed to evaluate the feasibility of source-separated collection for recycling. . . . This project has provided assistance to the advancement of environmentally sound recycling methods. This project continues as a service to the community and for its research potential."[13] While the numbers were not particularly impressive—by 1977 an estimated four thousand residents were participating—McGinnis was generally pleased with the results. As he notes, "We didn't have professional equipment. We didn't have blue boxes. Everybody had to use cardboard boxes or whatever. So there were definitely limits. What went well was the community involvement and the fact that people would listen to reason. People were proving what we believed in: people were naturally good, you just needed to give them the tools."[14] The IFF would later find out that theirs was the first roadside, multi-material pickup to operate in Canada.[15]

EXPANSION OF THE IS FIVE FOUNDATION

McGinnis's astute business sense enabled the IFF to expand dramatically in its second year. Seeking support from the Local Initiatives Program (LIP), a federal employment scheme, he recognized that there would be major competition for funding, which was capped at $100,000.

> We knew we were up against a lot of competition after our first year because other people had heard about the program and even though we'd done fairly well and they seemed to like what we'd done in year one we knew we'd have to be clever. And we wanted to get bigger and figured out they gave out the money riding by riding. So there was competition ... [within] a federal riding, but often there was a bit of money left over once they got done deciding who was going to get the priority. So we figured out how to come up with the smallest grant we could apply for—the least amount of people for the shortest amount of time. I did twenty-one applications, photocopied exactly the same with every federal riding in Toronto, except the one in the Beaches where we had our original grant. So with the Beaches we got another round of seven people as the head office, and out of the twenty-one [applications] we submitted they approved eleven of them, without knowing it. When they had their first get-togethers for the project officers to meet their new grantees, it was only then that they figured out how much money they'd give [laughs], which was well over $100,000.[16]

McGinnis' canny manoeuvring led to a revamped application process the following year, as LIP applicants were required to identify whether they were simultaneously applying for funding in any other federal ridings. Nonetheless, the LIP funds enabled the IFF to undertake a variety of projects, employing twenty-nine people full-time at its peak.

The bulk of the IFF's income, not to mention its public renown, came from its work in recycling; however, this was far from its only focus. From the outset, the aim was to create an environment in which individuals would work together to pursue their collective interests. A review of the group's periodical, the humorously named *Another Newsletter*, reveals a great diversity of projects that reflected the IFF's countercultural basis. These projects generally fell into one of three categories. The promotion of a healthy lifestyle was a major focus. The newsletter featured numerous easy-to-make yet healthy recipes. IFF members conducted a survey of the nutritional value of food options available to office workers taking their lunch break in downtown Toronto. They created an exercise booklet summarizing the advice of experts, noting that a fit body was essential to achieving physical and mental health. Likewise, in a September 1977 column, member Tim Michael provided a first-hand account of how he had managed to quit smoking, complete with practical tips.[17]

Energy issues were also prominent. While some attention was devoted to alternative energy sources, such as solar power, the subject of energy conservation was of particular interest. This can be seen in the inclusion of workshops and practical tips to help save on heating along with a demonstration of how old newspapers could be used as insulation.[18] The newsletter also revealed the organization's abiding interest in waste reduction. In the September 1977 issue, Michael Johnson explained how he recovered useful items such as furniture, an eight-track player, and Pirelli radial tires from the garbage. The group hosted a weekly flea market in its centrally located Dupont Street headquarters, with the proceeds helping to fund its activities. It also established a "community waste exchange" to redistribute useful but low-value items that would otherwise be destined for landfill.[19]

While environmentalists have generally advocated for the development of recycling programs, it is important to note that recycling was not viewed as an ecological panacea. In itself, recycling is not a particularly radical action, as it allows for the continuation of the consumer-driven lifestyle. According to the 3Rs waste hierarchy introduced in 1973 by the Toronto-based Pollution Probe, the key to combatting waste is threefold. First, individuals must reduce their consumption, as this will cut down on the wasteful use of raw materials and energy. Second, individuals must reuse items. Finally, those items that cannot be reused should be recycled. While recycling proved preferable to both sending products to landfill and utilizing virgin resources, it was the final step in this hierarchical process because extensive energy and resources are required to collect and physically recycle the materials. If the 3Rs are viewed as an inverse pyramid, "reduce" limits the amount of goods consumed, "reuse" further limits this, and "recycle" is only for the remaining items.[20] This philosophy was consistent with the thought process at the IFF. As was noted in an *Another Newsletter* article timed to coincide with the 1977 Christmas shopping season, "Recycling is good, but reduction is better. . . . Our first opportunity to limit our personal wastage of valuable natural resources generally comes as we are deciding what products to buy."[21] As Derek Stephenson noted in a 1978 interview with *Globe and Mail* reporter John Marshall, active participation in the recycling process was an important step in recognizing broader issues. He explained, "Individuals just can't see how they can clean up the Great Lakes, save the seals, stop rip-offs. But they can peel labels off cans. It's a start towards an acceptance of the environmental ethics of a conserver society."[22]

Stephenson was drawn to the success of the IFF. A recent graduate of the University of Western Ontario, he had studied under William Bunge, the innovative urban geographer and spatial theorist whose radical politics led him to flee the United States in the early 1970s.[23] "He knocked me right out of the system," Stephenson recalls.[24] Upon graduating, Stephenson and his classmate Tom Scanlan moved to Toronto to run, with six others, the Toronto Geographical Expedition. Patterned after the earlier, Bunge-led Detroit Geographical Expedition and Institute, which brought together locals and geographers from Michigan State University to study power dynamics within the African-American neighbourhood of Fitzgerald, the Toronto project brought together eight "urban explorers" who spent the year living in a house on Brunswick Street. Maintaining a constant dialogue with the local residents, they engaged in a sophisticated power analysis of the inner city, focused upon traffic patterns and park life as well as the effects of high-rise apartments on the development of children.[25] With this project completed, Stephenson and Scanlan backpacked throughout Europe, speaking at numerous universities about their work. Upon returning to Toronto, and looking for new projects to undertake, Stephenson was directed to McGinnis. As Stephenson recalls of their meeting,

> He and I hit it off, probably in the first fifteen minutes. He was describing all the things his organization wanted to do with public transportation, which was a strong interest of mine as an urban geographer, organic gardening, health-related things, energy conservation, and something called "recycling." . . . I said, "Well, this sounds very interesting. I would join you as the research director for Is Five but the one thing I'm not interested in is recycling" [be-] cause I said it didn't turn my crank and I don't see people sorting out their garbage, so how about I do everything else but that? And from that sort-of-fateful moment I started to get involved, getting introduced to the concept of recycling, and got drawn into it.[26]

Scanlan, incidentally, would also join IFF, focusing on the publication of educational textbooks and workbooks.[27]

As the IFF further established itself at the forefront of the local recycling industry, it began to attract something rare: volunteers with experience in the field. Such is the case of Toni Ellis. While studying at New Brunswick's Mount Allison University, Ellis had started a campus recycling program. Bothered by the amount of waste paper created by faculty, staff, and students, she filled her knapsack with an assortment of paper types and hitchhiked to Moncton, where she visited a recycling plant. Having determined that the plant could indeed make use of these papers, Ellis organized a collection program. Upon graduating in 1976 she moved to Ottawa, where she helped the local Pollution Probe affiliate establish a fine-paper recycling program, collecting materials from government offices in the city. A few months later she moved to Toronto and enrolled briefly in a public health inspection program at Ryerson Polytechnical Institute. After speaking to a city employee about her interest in recycling, Ellis was encouraged to contact Jack McGinnis; she subsequently visited the IFF office on Dupont Street and offered to volunteer. She laughs as she recalls that the response to her offer was not quite what she had expected: "They said, 'Oh yeah, sure, we could use some volunteers. Here, take the vacuum.'"[28] Ellis became one of dozens of volunteers dedicated to working with the IFF.

In 1978 the IFF launched two recycling programs in the Greater Toronto Area. Even with careful planning, the recycling industry was notoriously turbulent. This would become particularly evident in the case of a weekly newspaper-pickup program started in North York. Focusing on the area between Victoria Park Avenue and Bayview Avenue, and from Highway 401 to the borough's southern limit, the project was suspended shortly after it began, as the IFF's paper broker, Attic Insulation, went bankrupt.[29] However, this failure was offset by resounding success elsewhere. On December 8, 1977, the IFF submitted a proposal to the East York Works Committee to operate a weekly newspaper pickup throughout the borough, with IFF assuming all costs. Approved by the works committee four days later, the plant received the go-ahead from East York council on December 19, 1977.[30] Operating under the auspices of the East York Conservation Centre (EYCC), pickup began in February 1978, utilizing two trucks. Six months later the program had achieved 33 percent participation, averaging twenty-five to thirty tons of newspaper per week. By June 1979 these figures had increased to a 45 percent participation rate and thirty-five tons per week.[31] At this point the EYCC boasted of running "Canada's largest non-municipal source separate waste reclamation program" and expanded the program's scope to include collection of cardboard, glass, and metals.[32]

As explained in a November 1979 report, the "East York recycling project was initiated to provide a demonstration of the viability of local at-source recovery programs."[33] Documenting their extensive planning in a series of reports, the IFF also took the opportunity to study the functionality of various technologies and approaches to recycling. As the group acknowledged, a "major barrier to the successful implementation of at-source recovery on a broad scale was identified as a lack of suitable collection equipment designed for multimaterial curbside collection of recyclable materials."[34] Having started with a pickup truck in the Beaches in 1975, by the time the East York pickup began the IFF had purchased a GMC MagnaVan with a 2.5-ton carrying capacity and rented a similarly equipped vehicle. The foundation received funding through Environment Canada's Development and Demonstration of Resource and Energy Conservation Technology Program, which allowed them to collaborate with the Toronto-based DEL Equipment Ltd. in the creation of a vehicle specially designed for recycling programs. The resulting prototype cut down on the physical labour involved in collection, enabled a two-person crew to collect multiple waste streams, was capable of automatic unloading, and was priced competitively with existing collection vehicles. Having organized and run some of Toronto's pioneering recycling programs, the IFF was also intimately involved in the development of the associated recycling technology.

THE SPINOFF ORGANIZATIONS

The twin problems of rising unemployment levels and inflation in the aftermath of the 1973 energy crisis led the federal government to introduce austerity measures. This, Dennis Guest has noted, resulted in the cancellation of various governmental make-work programs, while LIP saw its budget cut dramatically beginning in 1976.[35] The IFF therefore had to seek alternative sources of income. This search was met with varying degrees of success. In addition to recycling, the group generated income from its flea markets and the use of its press, which printed materials for local non-profit organizations. Nonetheless, staff were forced to fund their work with personal savings and income drawn from elsewhere while they awaited results of new funding applications.[36] However, the solution to the IFF's financial woes soon appeared. As Stephenson explains, "We were starting to get lots of consultants, people in really nice suits, coming by our operation to learn how we were doing things. We would tell everybody everything. And it dawned on me sometime that we were providing information that consultants were then selling to clients for a lot of money. I thought, 'Wait a minute here, why don't we do the consulting?'"[37] In March 1977, Resource Integration Systems Ltd. (RIS) was launched to provide "consulting service in the field of conservation, with a particular emphasis on waste management and recovery systems."[38] With Stephenson serving as president, RIS began funding the IFF's activities by charging consultants' rates for its expertise.

In July 1977 RIS received a subcontract to design and implement a multi-material recycling program for Canadian Forces Base (CFB) Borden. This project was the brainchild of Rick Findlay, senior project engineer at Environment Canada's Environmental Protection Service, and had been inspired by a visit to the Centre for Resource Recovery (CRR), still under construction in the Downsview district of North York. The CRR was a mechanical separation system designed to handle all forms of recyclable materials. While the project was based on unproven technology, the province had invested $20 million in it. Convinced that separation at source would prove much more efficient than the unproven mechanical separation system in which the province had invested millions, Findlay chose CFB Borden because of its proximity to markets for recovered materials, the detailed knowledge of its past waste generation and management practices, and the willingness of the Department of National Defence to consent to the project.[39] This project meshed with the IFF's belief that separation at source was essential to environmental change, as it forced participants to consider their consumer habits. Stephenson recalls, "We were essentially given this place to experiment with recycling. Had a good budget, but we were subcontractors to consultants who were theoretical, MBA-types, while we were operational types. And from that experience we . . . got to play around with other people's money and perfected a lot of techniques."[40] The project resulted in the collection of corrugated boxes from shopping centres on the base, glass and bottles from its drinking establishments, paper and newsprint from its offices, and cans, newspaper, and glass from its residences. When it ended in March 1979, the project was considered a success, with 45.9 percent participation in the curbside collection of newspaper and 21.4 percent in the collection of glass.[41] It was subsequently determined that this program, if continued, could provide upwards of $15,000 in net profit annually.[42]

In 1978 Jack McGinnis secured a grant to spend three months meeting with recycling advocates and practitioners throughout Ontario in order to determine the need for a province-wide recycling organization. He also paid a visit to the West Coast in order to examine the model of the British Columbia Recycling Council, formed in 1973. As McGinnis later recalled, the trip left him with an unequivocal reaction: "For the first half of the tour . . . I'd tell people that we were thinking of forming a province-wide group. Halfway through, I was saying, 'We've formed a group.'"[43] For two days in June 1978 over one hundred interested parties gathered at the Holy Trinity Church in downtown Toronto to launch the Recycling Council of Ontario (RCO).[44]

Beginning its life in the IFF offices at 477 Dupont Street, the RCO had a twofold agenda: to serve as a network for the province's non-profit recycling groups, and to develop co-operative marketing for its members. It had an early brush with success when the Ontario Paper Company announced its decision to build a de-inking plant in Thorold. The RCO had offered to provide 64 percent of the plant's needs within three years; however, an unstable market and pressure from the province's traditional paper companies that now viewed the organization as a threat led the RCO to abandon its marketing efforts. Despite this, the RCO flourished as an information provider. In March 1981 it established the Ontario Recycling Information Service (ORIS), which created a toll-free telephone line to answer the public's queries about recycling and available programs. Modelled after a service operating in Portland, Oregon, ORIS was fielding 20,500 questions per year by 1990.[45]

The first executive director of the RCO was Eric Hellman. About to enter his freshman year at the University of Toronto with the intention of becoming an engineer, Hellman underwent what he described as "an epiphany" while visiting Manitoulin Island during the summer of 1974. Feeling a sense of "oneness" with Lake Huron and the sun, Hellman recalls, "I just fell into the beauty of it. I remember the inner conversation going something like 'I love this so much' and then a ... deep inner-voice said, 'Well, why don't you do something to help it?' Totally nothing I've ever experienced before, but this kind of a deeper, larger sense of self. In that moment I knew that I would be going into environmental work."[46] He promptly changed his course of study to reflect his newfound passion, supplementing it with volunteer experience at Pollution Probe, the city's preeminent environmentalist organization. Hellman later transferred to the University of Waterloo in order to pursue a degree in environmental studies. Hellman organized Garbage Fest 77, an event held on November 19, 1977, that brought many of the province's foremost environmentalists together in Waterloo to discuss waste issues. Having impressed members of the IFF, Hellman was hired to join RIS as a consultant before subsequently assuming his position with the RCO.[47]

BIRTH OF THE BLUE BOX

Garbage Fest 77 also brought the IFF into contact with Nyle Ludolph. The director of special projects at the waste management company Superior Sanitation, Ludolph had previously cared little for recycling. However, the day spent in the company of recycling advocates had a transformative effect upon him: "My conscience got a hold of me and I said, 'I'm going to try this.' I went home that day and dug up a hole in the backyard for compost, and I put boxes at the side door in the garage and I said to the family, 'We're going to test this recycling thing.' Consequently, we ... only generated 102 pounds of garbage for the entire year."[48] This amazed Ludolph, who notes that the average family of three at that time would normally generate a ton of garbage annually. As acquiring land for landfill sites was becoming increasingly difficult, he saw recycling as a way to help the company while at the same time earning the public's support. His boss, Ron

Murray, president of Laidlaw Waste Systems Ltd., was also intrigued with the potential; however, Murray worried about the potential business implications. As Ludolph recalls, "He said, 'Look, if we do that we may as well park the garbage trucks.' And I said, 'No, no. For every garbage truck we take off we put on a recycling truck. What's the difference?' He kind of agreed with that concept. We weren't going to hurt our business any—it would complement our business."[49] Despite Ludolph's optimism for the initiative, he admits that "A lot of garbage handlers thought we had lost our minds."[50]

Following RIS's success at CFB Borden, Ludolph approached Hellman about bringing recycling to Kitchener. According to Hellman,

> He said to me, 'Wouldn't it be amazing if we could do this city-wide? If everybody would do this?' And I'm looking at this guy who was head of garbage collection for this company going, 'Do I hear what I'm hearing? Does he actually want to do recycling?' I said, 'Now, if you're serious I'll give you a proposal.' So I went back to the office in Toronto that day and put together a proposal for the test program, which was approved by Superior [Sanitation] and became the foundation for the blue box.[51]

Hellman recalls Murray's response to the proposal:

> In the conversation about the proposal we had made to them he [Murray] said something very frank. 'We make our money off of garbage. We make a good living. But something in me says this can't last forever, that it doesn't make sense, business-wise or social-wise, to be paying somebody to keep picking up garbage. At some point this has to turn into something like recycling, where there's some good being made out of this material.'[52]

Hellman's proposal to examine the efficiency of a variety of collection methods from a sampling of one thousand homes in Kitchener received funding from Laidlaw. RIS was given the opportunity to design the project, which would be carried out by a new division of Laidlaw headed by Ludolph. The project was an astounding success. Originally scheduled for six months, beginning in September 1981, it continued uninterrupted until 1983, when the recycling program went city-wide. Particularly positive results emerged from the homes—a quarter of the total sample—given a blue box in which to place their recyclables. According to Stephenson,

> [for] the first hundred boxes that went out there, . . . participation rates went from one-third to maybe one-half of all households to essentially one hundred percent of all households. You gave them a box and people loved it. In fact, they loved it so much . . . we would get calls from different parts of Kitchener that would say, "You haven't picked up my blue box today." Well, they'd gone over to that neighbourhood and stolen one out of the test area and taken it over to their home.[53]

And why was blue chosen as the colour of the boxes? As Stephenson recalls,

> When we had the Kitchener program we were able to experiment with a hand-assembled one, what we used to call chloraplastics, and we assembled about 150 of these boxes. We hand stenciled them with "We Recycle." They happened to be blue . . . [because] with plastics the darker it is the less likely it will break down with ultraviolet light, at least in those days. We thought black was good for that, and black would stand out in the snow, but it wasn't very attractive. We didn't want to go the conventional green, and so we picked a spectrum in there that was our best guess for what the right color was. We picked blue.[54]

In 1983 Laidlaw's blue box program went city-wide in Kitchener. Almost immediately, participation levels hit 85 percent.[55] As Ludolph recalls, implementation of the program, which was strictly voluntary, was very easy. Bins, containing educational information, were left at the entrance of each home in the city. "When we distributed the 35,000 [blue boxes] I only had four people that said, 'Come take this thing away, we're not going to do this.' I must tell you that within a week three of these people called back and said they had changed their mind."[56] Despite the popularity of the expanded program, in which Laidlaw was heavily invested, it was nearly abandoned the following year when the company's contract with the city expired. While the company attempted to recoup some of its costs in its follow-up bid, it was revealed that a rival garbage contractor without a recycling plan had submitted a bid $400,000 lower than that of Laidlaw. However, at the ensuing general council meeting, public support for the blue box program, coupled with supportive presentations from Ludolph, Paul Taylor of the RCO, Pollution Probe's executive director Colin Isaacs, and a group of schoolchildren who recited a poem on the merits of recycling persuaded council to accept the higher bid.[57]

The blue box program continued to expand. In 1985 Laidlaw brought it to Mississauga. That same year, the Ontario Soft Drink Association (OSDA) made a deal with the provincial government: the Environmental Protection Act would be amended to allow the introduction to the Ontario market of non-refillable, but recyclable, aluminum and plastic containers. In return, the OSDA promised it would be recycling 50 percent of its containers by December 1988.[58] In 1986 the provincial government, industry, and municipalities struck an agreement to share the capital costs of starting a province-wide blue box recycling program. By 2011, 95 percent of Ontario households were serviced by blue box recycling programs.[59]

CONCLUSION

The Is Five Foundation ceased operations in the mid-1980s, although its name, or at least a close approximation, continues in the guise of Scanlan's Is Five Press. Meanwhile, the foundation's consultation arm, RIS, parlayed its recycling expertise into global expansion. By the late

1990s the firm had grown to encompass five offices across Canada, the United States, and Europe, with seventy-five employees and $8 million in annual sales. Stephenson notes with pride that RIS played an important role in setting up recycling programs in such far-flung locations as Belgium, France, and the United Kingdom. Such was the reputation and reach of RIS that it was acquired in 1998 by a British venture capital company, Enviros.[60]

Stephenson continues to work in the business of recycling consultation. Whatever his original misgivings were concerning recycling, as expressed to Jack McGinnis at the time of their first meeting, the industry has made him a wealthy man. He is currently the Director of Global Solutions at Reclay Group, an international waste management consultancy, and a member of the board at LRS Consulting Ltd., a London-based firm specializing in sustainable resource management.[61] McGinnis, meanwhile, left RIS in the 1980s and turned his focus to opening the Durham Conservation Centre in Pickering, which continues to operate as Durham SustainAbility, a non-profit environmental organization dedicated to educating and supplying the public with the tools necessary to live sustainable lives. He also established himself as a leader in operating recycling initiatives at special events, including the 1991 International Special Olympics, the 1992 Super Bowl, the 1996 Summer Olympics, and the 2002 Winter Olympics. On January 29, 2011, McGinnis passed away from respiratory failure. The ensuing accolades, published in the *Globe and Mail* and the *Toronto Star* as well as in various industry publications, highlighted his role as the "father of the blue box."[62]

Recycling proved to be a highly successful undertaking for McGinnis, Stephenson, and their colleagues. At the same time, it is interesting to note that the success of recycling led to the transformation of their *modi operandi*. RIS, after all, had been formed to help fund the operations of the countercultural IFF. While RIS initially attempted to maintain the spirit of its predecessor, the nature of its work rendered this impossible. Toni Ellis related a telling incident when officials from Alcan were scheduled to visit the IFF-RIS headquarters. Employees tended to dress casually in the office; however, with important figures set to visit, there was a frantic effort to look more businesslike. As Ellis recalls with a laugh, "Somebody ran around the office handing out ties to all the guys that worked there so they could suddenly look more legitimate."[63] The addition of clip-on ties may seem minor, but it underscores an important transformation. In order to conduct business, the staff at RIS were forced to professionalize. This meant dressing appropriately in order to be taken seriously; it also meant adopting standard office procedures and implements, such as personal computers.

This maturation of RIS coincided with the maturation of its staff. In the late 1970s they were, by and large, fresh out of university. By the early 1980s they had transitioned into working adults. According to Stephenson, "Not-for-profits [such as the IFF] are wonderful environments for young, motivated people for a period of their life, but if you can't give them a career and adequate amount of money they eventually drift away."[64] This is not to suggest that the group became entirely corporate in its culture. As Ellis explains, "While we were definitely getting more businesslike I don't think we were losing our drive or our common vision." Given their close working relationships and frequent social gatherings, she points out, RIS continued to feel much like a collective.[65]

The IFF was born out of McGinnis's desire to work with others in a fulfilling manner. While it would undertake a broad spectrum of activities, its greatest success came in the realm of recycling. In addition to operating some of Toronto's earliest recycling programs, the IFF helped build the provincial recycling infrastructure through the creation of the RCO. It would also create RIS, the recycling consulting firm that created the iconic blue box and later took its expertise global. The countercultural organization with the strange name left a lasting legacy.

1   The blue box is virtually synonymous with recycling in Ontario. A recent survey shows that 89 percent of Ontarians "feel that the Blue Box Program still remains the main driver of their recycling habits." Further demonstrating the significance of the program is the fact that 75 percent stated that "the Blue Box is their primary environmental effort." "2010 Stewardship Ontario Annual Report," Stewardship Ontario, accessed 5 September 2012, http://stewardshipontario.ca/wp-content/uploads/2013/03/SO_2010_Annual_Report_FINAL.pdf.

2   Tony Wong, "Ontario Blue Box Program Honored by UN Award," *Toronto Star* (hereafter *Star*), 15 September 1989, A7.

3   Stuart Henderson, *Making the Scene: Yorkville and Hip Toronto in the 1960s* (Toronto: University of Toronto Press, 2011).

4   Grant Goodbrand, *Therafields: The Rise and Fall of Lea Hindley-Smith's Psychoanalytic Commune* (Toronto: ECW Press, 2010).

5   Walter Isaacson, *The Innovators: How a Group of Hackers, Geniuses, and Geeks Created the Digital Revolution* (New York: Simon & Schuster, 2014), 264–76.

6   These background details were derived from conversations with McGinnis and his sister Pat. Jack McGinnis, interview with the author, 8 July 2008; Pat Roaderick, interview with the author, 13 November 2012.

7   McGinnis, interview; Roaderick, interview.

8   McGinnis, interview.

9   McGinnis, interview. The name is also described in "Is Five Foundation Research and Public Education," insert in *Another Newsletter* 2 (September 1977); John Marshall, "Metro Alchemists Turn Garbage into Gold," *Globe and Mail*, 1 March 1978, 5.

10   McGinnis, interview.

11   Ryan O'Connor, *The First Green Wave: Pollution Probe and the Origins of Environmental Activism in Ontario* (Vancouver: UBC Press, 2015), 88–89, 148–49.

12   The May 1972 edition of the *Probe Newsletter* listed information for twenty-four depots scattered throughout the Metropolitan area, while a May 19, 1972, profile of Metro recycling efforts in the *Star* estimated the number was closer to forty. These varied greatly, from community-oriented endeavours such as the Newtonbrook Secondary School depot, which accepted newspapers between noon and four o'clock on Sundays, and the Boy Scouts of Pack 33 depot, which operated out of a private residence on Christie Street, to for-profit operations such as the Canadian Paper Fibres Company on Commissioner Street and Consumers Glass Company on Kipling Avenue. "Metro

Recycling Depots," insert in *Probe Newsletter* 4, no. 3 (1 May 1972); E. H. Hausmann, "Waste Recycling Efforts Only Scratch the Surface," *Star*, 19 May 1972, 8.

13   "Is Five Foundation Research and Public Education."

14   McGinnis, interview.

15   Ibid.; Diane Humphries, *We Recycle: The Creators of the Blue Box Programme* (Toronto: Pollution Probe, 1997), p. 6, accessed 7 March 2008, http://www.pollutionprobe.org/Reports/we%20recycle.pdf.

16   McGinnis, interview. The actual amount, according to a contemporary newspaper report, was just over $129,000. Jacques Bendavid, "Environmental Group Awaits New Funding," *Star*, 1 September 1976, F3.

17   "Food and Nutrition," *Another Newsletter* 2 (October 1977): 5; Tim Michael, "Is Five Fitness," *Another Newsletter* 1 (June–July 1977): 6; Tim Michael, "Let's Quit Smoking," *Another Newsletter* 2 (September 1977): 6.

18   *Another Newsletter* printed an educational overview of solar energy, complete with a reading list for further information, while Bob Argue, vice-president of the Solar Energy Society of Canada, gave a presentation on the subject during the IFF's March 1977 general meeting. "Solar Energy," *Another Newsletter* 1 (March 1977): 2–3; "General Meeting," *Another Newsletter* 1 (March 1977): 3.

For information on the group's conservation efforts, see Arthur Jacobs, "Energy Conservation—A New Frontier," *Another Newsletter* 1 (June–July 1977): 5; "The Is Five Seminars and Practical Workshops," *Another Newsletter* 2 (October 1977): 2; "Energy Conservation: Making Insulation from Old Newspapers," *Another Newsletter* 2 (October 1977): insert.

19   Michael Johnson, "Junkie!" *Another Newsletter* 2 (September 1977): 7; "Warehouse," *Another Newsletter* 1 (February 1977): 5; Jack McGinnis, "Is Five Inaugurates Free Waste Exchange," *Another Newsletter* 1 (June–July 1977), 2.

20   Gregory Bryce to the Toronto Recycling Action Committee, 10 January 1973, F1057 MU7361, Ontario Archives, Toronto; O'Connor, *The First Green Wave*, 110–12.

21   "Reduction is Better," *Another Newsletter* 2 (November 1977), 6.

22   Marshall, "Metro Alchemists," 5.

23   Bunge was an unrepentant Communist who appeared between H. Rap Brown and Stokely Carmichael on a 1970 House Internal Security Committee list of radical campus speakers. David E. Rosenbaum, "House Panel Lists 'Radical' Speakers," *New York Times*, 15 October 1970, 23.

24   Derek Stephenson, interview with the author, 11 December 2009.

25 For more information about these projects, see Ronald J. Horvath, "The 'Detroit Geographical Expedition and Institute' Experience," *Antipode* 3, no. 1 (1971): 73–85; Derek Stephenson, "The Toronto Geographical Expedition," *Antipode* 6, no. 2 (1974): 98–101; Clark Akatiff, "'Then, like now . . .': The Roots of Radical Geography, a Personal Account," *Antipode*, posted 4 September 2012, http://antipodefoundation.org/2012/09/04/then-like-now-the-roots-of-radical-geography-a-personal-account.

26 Stephenson, interview.

27 Tom Scanlan, "Is Five and Education," *Another Newsletter* 1 (June–July 1977): 1–2; Tom Scanlan, interview with the author, 10 June 2011.

28 Toni Ellis, interview with the author, 26 October 2012.

29 Harold Hilliard, "North York Okays Newspaper Pick-Up for 3-Month Trial," *Star*, 5 September 1979, A22.

30 "Environment/Jobs—Conflict or Harmony," *Another Newsletter* 2 (December 1977): 2.

31 *Investigation of the Feasibility of Increasing Corrugate Cardboard Recovery through Industrial and Commercial Source Separation in Ontario* (Toronto: Is Five Foundation/Resource Integration Systems, 1979), 1.

32 Ibid., i (quotation), 8.

33 *Description and Evaluation of the East York Recycling Model* (Toronto: Resources Integration Systems, 1979), 7.

34 Is Five Foundation, *Development and Demonstration of a Customized Truck for Collection of Glass, Metal and Paper Refuse* (Ottawa: Environment Canada, Technical Services Branch, 1983), 1.

35 Dennis Guest, *The Emergence of Social Security in Canada*, 3rd ed. (Vancouver: UBC Press, 2003), 180–81.

36 "Self-Support Survival," *Another Newsletter* 2 (November 1977): 2–3; Marshall, "Metro Alchemists," 5.

37 Stephenson, interview.

38 Resource Integration Systems (hereafter RIS), *Submission to the Ontario Paper Company* (Toronto: 1980), 2.

39 RIS, *At-Source Recovery of Waste Materials from CFB Borden: The Viability of At-Source Recovery in Small Communities, Executive Summary* (Toronto: RIS, 1979), 1.

40 Stephenson, interview. CFB Borden consisted of 1,276 residences and nine commercial operations. The average weekly pickup totalled 4,515 pounds of newspaper and 907 pounds of glass from the residences, alongside 2,000 pounds of corrugated boxes, 440 pounds of glass, 1,000 pounds of computer paper, and 958 pounds of ledger-grade paper from the commercial operations and offices. RIS, *At-Source Recovery*, 33–34.

41 Ibid., 33.

42 Ibid., 7.

43 Quoted in Katharine Partridge, "RCO Celebrates 20 Years!" *RCO Update*, October 1998, 2.

44 Ibid., 1–2; "Government Policies Thwarting Recycling, Conference to be Told," *Globe and Mail*, 1 June 1978, 3; McGinnis, interview.

45 The toll-free service, renamed the Waste Reduction Information Service in 1993, was discontinued in 1996 as a result of government cutbacks. However, as of 1998 a walk-in service continued to operate. Partridge, "RCO Celebrates," 3, 5–7; Eric Hellman, interview with the author, 12 January 2010.

46 Hellman, interview.

47 "Garbage Fest 77," *Another Newsletter* 2 (October 1977): 6; Hellman, interview.

48 Nyle Ludolph, interview with the author, 16 January 2010. While recycling was seen as an economically viable activity, it appears that composting did not inspire the same confidence. Subsequently, compost pickups have historically been much scarcer in Ontario.

49 Ibid.

50 Ibid.

51 Hellman, interview.

52 Ibid. As Ludolph recalls of the meeting, "Eric had said, 'You're a waste management company but you don't manage waste. You just pick it up and bury it.' [*laughs*] My president, Ron Murray, was impressed with the honesty of Eric Hellman. He was very impressed, so he gave him an ear, and he agreed to give him $70,000 in an experimental program in recycling." Ludolph, interview.

53 Stephenson, interview.

54 Ibid. Eric Hellman tells a slightly different story: "Jack [McGinnis] was the one who went to the plastics manufacturer and was looking at what kind of boxes we could get, found one that was reasonably economical and that was blue. It was a plastic corrugated container and that became the reason why it was blue." Hellman, interview.

55 Humphries, *We Recycle*, 8.

56 Ludolph, interview.

57 Ibid.; Stephenson, interview; Humphries, *We Recycle*, 8.

58 *A Brief History of Waste Diversion in Ontario* (Toronto: Canadian Institute for Environmental Law and Policy, 2008), 2; David McRobert, "Ontario's Blue Box System: A Case Study of Government's Role in the Technological Change Process, 1970–1991," (LLM thesis, York University, 1994), 40. While the blue box system would continue to expand across the province, dramatically increasing the recycling participation rate, the share of refillable soft drink containers would drop from a market share of 40 percent in 1986 to just 3 percent in 1993. Derek Ferguson, "NDP Record

on Refillables Criticized," *Star*, 24 November 1993, A10.

59 Stewardship Ontario, "The Story of Ontario's Blue Box," p. 19, accessed 15 September 2015, http://stewardshipontario.ca/wp-content/uploads/2013/02/Blue-Box-History-eBook-FINAL-022513.pdf.

60 Stephenson, interview; Geoff Love, "Curriculum Vitae," Love Environment Inc., accessed 10 September 2012, http://www.loveenvironment.com/cv.html.

61 "Derek Stephenson," LRS Consultancy website, accessed 10 September 2012, http://www.lrsconsultancy.com/about-us/our-team/3/Derek-Stephenson.

62 Louise Brown, "'Father of the Blue Box' Died This Week," *Star*, 4 February 2011, http://www.thestar.com/news/gta/article/933197---father-of-the-blue-box-died-this-week; Tony Leighton, "Jack McGinnis," *Globe and Mail*, 22 April 2011, http://v1.theglobeandmail.com/servlet/story/LAC.20110422.LFLIVES0422MCGINNIS-ATL/BDAStory/BDA/deaths; Guy Crittenden, "Recycling Council of Ontario and Blue Box Founder Passes Away on January 29, 2011," *Solid Waste & Recycling*, 1 February 2011, http://www.solidwastemag.com/news/jack-mcginnis-obituary/1000401159/; Karen Stephenson, "Jack McGinnis—King of the Curbside," *Green Solutions Magazine*, 1 April 2011, http://www.greensolutionsmag.com/?p=1765.

63 Ellis, interview.

64 Stephenson, interview.

65 Ellis, interview.

"Vive la Vélorution!": Le Monde à Bicyclette and the Origins of Cycling Advocacy in Montreal[1]

Daniel Ross

Montreal loves the bicycle. In 2013 the *Copenhagenize Urban Cycling Index* ranked it the most bike-friendly city in North America, and eleventh worldwide. On the island of Montreal, 36 percent of adults and 57 percent of children cycle at least once per week, and a hundred bike shops sell upwards of ninety thousand new bikes every year. Even more striking is the high level of support the City of Montreal has shown for urban cycling in recent years. As a result, the city boasts one of the largest networks of bike lanes and paths in North America—over six hundred kilometres—and a hugely popular system of five thousand public bicycles (BIXI) that, since its founding in 2009, has been exported to nine other cities around the world. While the car may remain king in North American cities, Montreal seems to be one of the places where the humble bicycle stands the best chance of challenging its reign.

But the status of the bicycle in Montreal has not always been so sunny. In the mid-1970s, Montreal cyclists were frustrated. Despite the growing popularity of cycling for transport, riding a bike on city streets was difficult and dangerous. At city hall, the Civic Party administration showed little interest in changing its pro-car stance to accommodate cyclists. In response, in 1975 a small group of cyclists banded together to found Canada's first major urban cycling advocacy organization, Le Monde à bicyclette (MAB). Drawing on a wide range of influences, including the 1960s and 1970s counterculture and the environmental and urban reform movements, the MAB adopted a distinctive mix of tactics and ideas centred on the social and environmental benefits of the bicycle. In choosing the bicycle as their form of resistance to car culture and consumer capitalism, the MAB grounded a countercultural critique in everyday practice, while adding an element of spectacle to its demonstrations. The group quickly became well known for its creative and provocative street theatre pieces (cyclo-dramas) and its calls for a *vélorution* ("*vélo*" being French for "bicycle") that would end the dominance of the private automobile. Operating within a growing network of cycling organizations in North America and Europe, the MAB were early proponents of what scholars have recently called "the cycling counterculture."[2] A closer look at the group's origins and activities sheds light on Montreal's vibrant culture of dissent in the 1970s and the local and international influences that helped to shape it. Today, the story of Le Monde à bicyclette is more relevant than ever: Montreal's current cycling renaissance owes a great deal to plans first put forward by the group, which many in the 1970s considered marginal, radical, or just plain crazy.

BIKE BOOM

A number of cycling scholars have described how North Americans went bike crazy in the last decade of the nineteenth century.[3] The founding of the MAB in 1975 came on the heels of another, equally dramatic, boom in popular interest in cycling. After a decade of steady growth, in the early 1970s sales and ridership increased dramatically in both Canada and the United States. Between 1970 and 1972, bicycle sales in the United States doubled to nearly fourteen million, and for the first time, bikes outsold cars; in Canada during the same period, sales reached a record two million.[4] A certain enthusiasm accompanied the rush to buy bikes. One Montreal newspaper, noting in 1972 that the city's bike shops had doubled their sales, exclaimed that it was "The return of the good old days!"[5]

Several factors seem to have contributed to this bike boom. Technological change was one, just as it had been in the 1890s. This time, however, it was not mass-produced pneumatic tires that made cycling more attractive, but the availability of cheap, lightweight, multi-speed bicycles, often imported from Europe or Asia. Demographics also mattered; by the 1970s the baby boom generation had graduated to adult bicycles, providing a huge market for the new technology. And from 1973 onwards, skyrocketing gas prices and fears of fuel scarcity caused by the oil crisis made the bicycle a more attractive alternative to the automobile.[6]

People also adopted the bicycle for the individual and social benefits it promised. The bike boom owed something to the growing popular interest in physical fitness and health that characterized the period and helped to move dieting and activities like bodybuilding, jogging, and aerobics from the margins of North American culture into the mainstream.[7] And, in an era of growing environmental consciousness, the bicycle was one of the greenest transport choices available. All of these factors contributed to a historic bike boom in the 1970s that put many new riders—including record numbers of adults—on the streets of North American cities.

PEDAL AT YOUR OWN RISK

In Montreal, as in other large North American cities, cyclists quickly discovered that the infrastructure needed to make cycling for transport safe and viable was sorely lacking. By 1975 several Quebec municipalities— including Longueuil, across the river—were experimenting with urban bike paths. However, despite having a high population density that made it ideal for cycling, the island of Montreal did not have a single bike lane or path. This forced riders into direct competition with drivers for street space, provoking accidents and confrontations. As Marc Raboy, a cyclist in the 1970s and one of the earliest members of the MAB, remembers, to bike for transport in the city meant "literally taking your life in your hands."[8] Across Quebec, an average of sixty-eight cyclists died on the road each year during the early 1970s, rising to a peak of eighty-four deaths in 1974—more than five times today's numbers, despite the doubling of ridership since the 1970s.[9] Montreal had no system of bicycle parking stands or posts, and with bikes barred from buses, the metro, and all but one bridge over the St. Lawrence River, there were precious few options for crossing between the island and the South Shore on two wheels.

These daily frustrations were compounded by the unresponsiveness of the municipal government on the issue. By the mid-1970s, Mayor Jean Drapeau and the Civic Party had been in power continuously for more than a decade. Years of centralization had concentrated decision-making power in the hands of the mayor and his inner circle, a group of men who saw the future of Montreal in terms of large-scale modernization projects, including highways, stadiums, shopping malls, and apartment towers. They were willing to invest in a costly Vélodrome as part of the Olympic Games complex (now the Montreal Biodome) but were dismissive of plans to create space for cycling on city streets. In this closed-door political culture, cycling did not have a lobby at city hall. Provincial cycling organization La Fédération québécoise de cyclotourisme concentrated its efforts on lobbying for rural touring routes and changes to provincial law. Independent attempts to promote the issue—such as that of traffic safety advocate Gilles Roger Prevost, who in 1972 presented the city with a report calling for an ambitious 2,400 kilometres of urban bike lanes—were met with inaction. When questioned on the subject in May 1975, the head of Montreal's traffic department, Jacques Barrière, summed up the dominant view among the city executive: "Bicycles are not a priority at the moment. If we encourage the bicycle too much, will we put the cars in our pockets?"[10] To attract attention to their needs, cyclists realized that they would have to organize outside of municipal politics.

FOUNDING A MOVEMENT

In this context of growing "cyclo-frustration" (a word coined by the group), Le Monde à bicyclette was founded. In April 1975, a small notice appeared in the *Montreal Star* announcing a meeting of a group tentatively called the Montreal Bicycle Movement. The notice went on to say that the movement had plans to "organize joyful cycling events" and "press city authorities for facilities." Anyone interested in cycling was welcome to attend.[11]

The announcement in the *Star* gave substance to informal discussions about cycling among a few young cyclists and activists based in the neighbourhoods just east of McGill University and Mount Royal: Milton-Park, Mile End, and the west Plateau. Those neighbourhoods were home to a vibrant mixture of French- and English-speakers, immigrants, workers, students, artists, and intellectuals. Their proximity to the downtown core and Montreal's universities, as well as the availability of low-cost rental units, made them key sites for the city's 1960s and 1970s counterculture; these areas were dotted with communes, co-ops, alternative bookstores, and art studios.[12] They were also hotbeds for political and community activism: Milton-Park, for example, was the site of a highly publicized confrontation between local residents and developers over plans to tear down several blocks of houses and replace them with an upscale residential/commercial complex.[13] In the 1974 municipal election, the St. Louis electoral district, which included the three neighbourhoods, elected three councillors from the Montreal Citizens' Movement (MCM), a coalition of progressive Montrealers formed to unseat Drapeau and renew Montreal's services and urban environment.

Over the next month a core of a dozen or so members began to meet on a weekly basis. As Raboy recalls, all but one of these early joiners were Anglophones, and meetings took place in English. Most of the group knew one another from involvement either in the 1974 MCM campaign or in local organizations like the St. Louis Health Food Co-op. The meetings took place in co-founder Robert

6.1 Thousands of cyclists take over Saint Catherine Street during the MAB Bike Week parade in 1976. Source: MAB Archives.

Silverman's apartment, much to the chagrin of his landlord, who complained about the pile of bicycles parked outside his front door.[14]

From the start, Silverman played an important role in the group. Older than most of the early members—he was forty-two in 1975— his life experience was wide and eclectic. Born in Montreal, he had (briefly) attended both English- and French-language universities,

DANIEL ROSS

had worked as a taxi driver and an English teacher, and for a short time had run an alternative bookstore. In the mid-1960s he had organized demonstrations against the Vietnam War with the Trotskyist Ligue socialiste ouvrière. With characteristic ideological commitment, since discovering the bicycle in France in 1969, Silverman has refused to drive a car. He was responsible for cultivating the MAB's connections with international cycling organizations and was one the group's main theoreticians and spokespersons.

From the initial meetings at Silverman's apartment came the group's name: Citizens on Cycles, in English, and Le Monde à bicyclette (meaning both "the people on bikes" and "everyone get on a bike!"), in French. They made plans to launch the organization city-wide at the end of May 1975 with a series of public events called Montreal Bicycle Week and the publication of a cyclist's manifesto. The MCM quickly lent its support to the idea, as did several other organizations; from early on, the MAB benefited from the willingness of other local environmental and community groups to work together on important campaigns.[15] Inspiration came from south of the border, too. On a visit to Washington, DC, in April, Silverman tapped into the network of cycling organizations that had sprung up in the eastern United States. His principal contact was John Dowlin of the Philadelphia Bicycle Coalition (PBC), a group that had been promoting urban cycling since 1971. The enthusiastic Dowlin sent Silverman a sheaf of material, including back issues of the PBC newsletter and news clippings about their activities.[16]

Montreal's first Bicycle Week, from May 26 to 31, 1975, was a dramatic success. Its popularity suddenly made urban cycling an issue in Montreal and was the impetus behind the MAB's transformation into a large, mass-membership organization. All of Montreal's major French and English newspapers commented on the MAB's "Bicyclist's Manifesto." The events organized by the group—the theatrical presentation of a bicycle to Montreal's city council; a commuter race between cyclists, drivers, and transit users; and a two-wheeled parade on May 31—captured the attention of the press and the public. Even the organizers were astonished by the three thousand cyclists that joined the Saturday afternoon parade as it rolled and shouted its way through downtown Montreal.[17] Less than two months after its founding meeting, the organization was on its way to becoming a fixture in public debates over transport and the environment in Montreal. By early 1977 the group had welcomed dozens of new members—membership would soon peak at around four hundred—and, thanks to a federal Local Initiatives Program (LIP) grant received in the winter of 1975/1976, had an office and its own newsletter dedicated to cycling culture and the environment.[18] As with many other activist organizations operating in the late 1960s and early 1970s, federal funding played a crucial role in sustaining the MAB.

With growth in its membership base, the composition of the MAB came to reflect not just the Plateau/Mile End area, but the city of Montreal as a whole. Of the dozens of environmental organizations founded in Quebec in the 1970s, the MAB was one of the most successful at acquiring new members.[19] Cyclists of different ages and walks of life joined from neighbourhoods across the city; most were Francophones. In changing from a small core of English-speakers to a larger group dominated by French-speakers, the MAB mirrored several other activist organizations formed in 1970s Montreal, including the MCM.[20] One new member was Claire Morissette, an environmentalist with an interest in cycling and other green technologies. While only twenty-five, she was already experienced in public outreach and active in the alternative St. Louis scene: she was a founding member of the Friends of the Montreal Botanical Garden and helped run the St. Louis Health Food Co-op, the first of its kind in Quebec. Morissette's organizational and literary skills, environmental vision, and creative energy would be tremendously influential within the MAB.

STOP THE JUGGERNAUT! VIVE LA VÉLORUTION!

From the start, the MAB combined a countercultural critique of mainstream society and culture with a willingness to pursue more modest, immediate goals. Two major themes ran through the group's ideology: the environmental and social destructiveness of the car, and the revolutionary potential of the bicycle. These ideas provided inspiration for the MAB's demands and turned riding a bicycle on city streets into a subversive act.

Le Monde à bicyclette saw the private automobile as a destructive force and the embodiment of the principal wrongs of Western society under capitalism: the alienation of the individual, the triumph of rationality and profit over well-being, and the systematic degradation of the environment. The MAB's publications are peppered with angry indictments of the damage done by the car:

> It destroys our homes, our green spaces, and our heritage, to build parking lots; it attacks our health, our lungs, our eardrums, our nervous systems; it empties our wallets and enslaves Quebecers [financially].[21]

The MAB focused in particular on deaths caused by cars and the damage cars did to the environment. For example, a press release protesting the 1976 Montreal Auto Show calls the car "public enemy number 1," adding that "since 1900, they have killed 25 million, more than were killed in all the wars of the 20th century. . . . It is our domestic Vietnam."[22] Meanwhile, the "Bicyclist's Manifesto" rails against cars for filling the air with "poisonous fumes," robbing the earth of raw materials, filling dumps with tons and tons of useless metal, and polluting the oceans with oil spills caused by their insatiable demand for fuel.[23]

Yet, to the MAB, the problem of the car is not just its social and environmental costs, but its cultural embeddedness. The radical ethos of the group drew on countercultural imagery that challenged the economic assumptions behind automobile use. Like the mythical

Hindu Juggernaut, the car is worshipped by those it destroys. The automobile has become so central to the dominant culture that people have ceased to recognize the possibility of an alternative. Car companies fill the airwaves with advertising and pressure governments into the "open squandering of millions" for highways and oil exploration. Owning and driving cars has alienated people from their bodies, their surroundings, and each other. Trapped behind steel and glass, drivers see pedestrians not as fellow human beings, but as obstacles. Furthermore, the car drives a wedge between the sexes, giving men control over women's mobility while the female body is used to market new killing machines. In sum, "[t]he automobile pollutes our values, tastes, ideals, in fact our very souls. It not only robs us of valuable raw materials, it steals our integrity as human beings."[24]

The MAB was not the first group to point to the private car as a symbol or agent of the ills of Western society; for many within the North American counterculture of the 1960s and 1970s, the car was a key component of a technocratic culture that threatened individual freedoms and human sensibilities. As early as 1961, social critics and urban theorists Paul Goodman and Percival Goodman published a serious plan for banning the car from the centre of New York City; from 1967, *Whole Earth Catalog* founder Stewart Brand and others called for a move from large, damaging machines and systems toward "appropriate technologies" on a human scale.[25] In Quebec, the ideas of the American and European counterculture were widely available in French through alternative media like the magazine *Mainmise*, which published seventy-eight issues between 1970 and 1978, as well as a French-language catalogue inspired by Brand.[26]

Likewise, urban reformers and environmentalists opposed both the car and the system that it represented. As historian Danielle Robinson has pointed out, urban dwellers across Canada organized in the 1960s and 1970s to oppose expressways in their cities, arguing much like the MAB that auto-centric planning destroyed urban communities and damaged the environment.[27] From the late 1960s onwards, air pollution was one of the main problems around which environmentalists in North America and Europe mobilized. In

6.2 MAB members in one of the group's largest die-ins, on Saint Catherine Street in 1976. The banner reads "Bike for life," and in the background a cyclist is being carried off on a stretcher. Source: MAB Archives.

Quebec, the first wave of the new environmental movement—more radical than its conservationist predecessors and unafraid to draw links between environmental and social problems—was spearheaded from 1970 by anti-pollution groups (and MAB allies) La Société pour vaincre la pollution and the Society to Overcome Pollution (STOP).[28]

What was novel about the MAB was the way it provided a single solution to the problems symbolized by the private automobile: the adoption of the bicycle. This position was heavily influenced by social theorist Ivan Illich, author of *Energy and Equity* (1974). Illich proposed an inverse relationship between the energy a society consumed and the equity of the distribution of wealth among its members: in his words, "high quanta of energy degrade social relations just as inevitably as they destroy the physical milieu."[29] On the other hand, "convivial" technologies such as the bicycle have a revolutionary potential to liberate the individual and create a more equitable society. This was the intellectual grounding of the *vélorution*.[30]

Compared to the car, the bicycle is cheap, accessible, and ecologically and socially harmonious. It is also a symbol of the possibility of a different way of life: "For that vehicle of death, we must substitute the vehicle of life: the BICYCLE. . . . [Our movement] endeavours to persuade those now dependent upon automobiles to become independent upon bicycles."[31]

For the MAB, choosing the bicycle liberates the individual in a way that riding the bus or subway—both green and equitable technologies compared to the car—do not. The cyclist moves at her own pace through the city, free from participating in the destruction caused by car capitalism. A critical mass of cyclists—the vélorution—would change social relations, politics, and humanity's relationship to the environment. Instead of society being divided into those who have cars—and control the streets—and those who do not, urban space would be equally accessible to all, under their own power. Government resources wasted on auto-centric development could be used for public transit and social services. As Silverman starkly put it, "the Juggernaut will die and we will all be better for it."[32]

The MAB was an intellectually heterogeneous organization, and the core themes of opposition to the car and promotion of the bicycle do not capture the full range of ideas held by its members. Members differed, for example, in where they saw the bicycle on the continuum between mode of transport and revolutionary tool and in the extent to which they linked cycling activism to other social movements. In March 1977, a founding convention was organized in an attempt to reconcile the different ideological orientations proposed for the MAB. As Claire Morissette remembers, this was

> no mean task, since [the choices were] to subordinate cy-
> clists' demands to the class struggle, to lobby from inside
> "the system," to preach by example by creating cycling

DANIEL ROSS

services, or to celebrate the bicycle while poking fun at the contradictions of the establishment![33]

While the "poetic-vélorutionary" position put forward by Morissette and Silverman had wide acceptance, there was never doctrinal unity in an organization that prided itself on being anti-authoritarian.[34]

Despite its big thinking, the MAB always brought at least a kernel of pragmatism to the table. The cycling counterculture it elaborated was accompanied by concrete demands aimed at ending the cyclo-frustration that had led to the group's founding.[35] From 1975 onwards, the MAB called for a bicycle commuting system made up of both physically separated north-south and east-west bikeways on major arteries and painted bike lanes on minor streets. They also demanded theft-resistant bicycle parking stands installed across the city and public education on cyclists' rights. To solve the problem of crossing the St. Lawrence River, they called for bike access to bridges and tunnels as well as to metro trains outside of rush hours. Recognizing that cycling was not ideal in all situations, the MAB argued that cycling improvements had to be accompanied by a massive expansion of Montreal's public transit system. Finally, the group envisioned a city-wide system of public bicycles:

> The City of Montreal must buy 10,000 bicycles and put them at the disposal, as community property, of the people of our city. So as to make them visible at night the city will paint them orange and for identification they will all be branded "M" and stamped with the seal of the city of Montreal. These bicycles would be kept in municipal storage centres throughout the city. To ensure that no antisocial person would steal community property, a deposit and identification would be required when taking out a bicycle.[36]

The MAB's demands were strongly influenced by ideas coming from other cities. While cycling advocacy was still in its infancy in other Canadian cities, allies like the PBC and New York City's Transportation Alternatives had been demanding cycling facilities since the early 1970s. Meanwhile, the "orange bicycle plan" was inspired by the European counterculture. In 1965, an eclectic Dutch group called the Provos (short for "Provotariat") presented a series of plans for improving Dutch society. Along with free birth control, shared parenting, legalization of squatting, and a tax on polluters, they drew up a "white bicycle plan" that they later attempted to put into action. The group painted fifty bicycles white and left them, unlocked, on the streets of Amsterdam for public use. Unfortunately, the Amsterdam police confiscated those bikes that had not been stolen. In 1967, Silverman met with several members of the Provos in Amsterdam—an experience that led directly to the MAB's championing of a similar, if more elaborate, plan for Montreal.[37] More than three decades before BIXI, MAB's 1975 "Bicyclist's Manifesto" marked one of the first appearances in North America of the idea of a public bicycle system.

These international influences reflected the fact that the MAB saw the vélorution as something larger than their own local struggle, consistently linking their own actions to the work of cycling organizations around the world. Every issue of the association's newsletter contained a section dedicated to international cycling news, and in 1978 the MAB was one of thirteen groups that founded the Cyclists' Internationale at a meeting in New York.[38] Also represented was the smaller Toronto City Cycling Committee (TCCC), the only other Canadian urban cycling group active at the time.

This sense of promoting a cause that transcended national borders may explain why Quebec nationalism was never a dominating factor within the MAB. The issue of language did arise early in the group's history: in 1977 the MAB moved from a haphazard bilingualism to adopting French as its official working language. While that decision was likely influenced in part by the rising tide of nationalism that had accompanied the Parti Québécois (PQ) victory provincially in 1976, it was above all a pragmatic choice that reflected both the new predominance of Francophones within the group and the willingness of Anglophone members to work in French. The change, like the MAB's limited support for the PQ government in Quebec City, does not seem to have caused much tension between linguistic groups, nor did it lead to an exodus of English-speaking members.[39]

DELIVERING THE MESSAGE: THE CYCLO-DRAMA

While the MAB intervened in formal politics—for example, endorsing the Montreal Citizens' Movement against the Civic Party—its members refused to express their demands through conventional channels. The mistrust that the MAB's radical core felt toward "politicians, bureaucrats, and other cocktail-lovers" (in Claire Morissette's words) was compounded by the group's complicated relationship with the MCM.[40] Initially the two were close; the founding of the MAB owed a great deal to the MCM's early commitment to encouraging grassroots activism, and the MCM's 1975 pro-cycling council motion was supported by an MAB rally on the steps of city hall. But relations cooled somewhat over the next few years. In the face of Drapeau's tight control of the city executive, reformers failed to achieve clear successes for cyclists. Additionally, the MCM's adoption of a more electoralist stance alienated grassroots groups like the MAB. It became clear that simply lobbying government was not an option.[41] Instead, the MAB's primary means of delivering its message became a provocative brand of street theatre called the cyclo-drama. Attention-grabbing, unexpected, and inexpensive to organize, the MAB's demonstrations spread their message to a wide audience and made them a staple of Montreal's oppositional culture in the 1970s.

One kind of cyclo-drama targeted specific cycling problems or demands. For example, in July 1978, after three years of equivocation by the city on the issue of bike lanes in central Montreal, the MAB took matters into their own hands. Overnight, two bidirectional lanes totalling just over two kilometres were painted on Saint Urbain and Marie-Anne streets, and motorists parked nearby received official-looking warnings from "Montréal: Ville cyclable" (Montreal: Bikeable City) calling on them to support the initiative. Journalists visiting the lanes—one of which was dubbed *Poumon rose* (pink

6.3 Guerilla lane-painting, likely 1980. Claire Morissette is in white. Source: MAB Archives.

lung)—were treated to an inauguration, complete with ribbon cutting, by MAB members.[42] This guerilla painting was repeated several times, including a 1980 episode in which Silverman and another MAB member were arrested with paint on their hands. They both eventually served a few days in Montreal's Bordeaux Prison, where Silverman recalls being comforted by his view of a nearby park's bike paths.[43] Today, Saint Urbain Street boasts a 2.5-kilometre bike lane that includes the portion briefly known as Poumon rose.

Many cyclo-dramas aimed at getting cyclists access to the Montreal metro, a battle that occupied much of the MAB's energies until the group's victory over the Montreal Transit Commission in court in 1983. Cyclists, often wearing gas masks or playing instruments, invaded the metro with their bikes and similarly sized (but permitted) objects like ironing boards, skis, and, in one case, a giant stuffed hippopotamus. Other demonstrations highlighted the absence of viable cycling links across the St. Lawrence: for example, at Easter

6.4 Parting the St. Lawrence River, 1981. Robert Silverman is holding the tablets. Source: MAB Archives.

in 1981 cyclists dressed in biblical-era costumes attempted to part the waters of the river to get across.[44]

A second category of street theatre aimed more broadly at "elevating consciences" by calling attention to the absurd contradictions of auto-centricity.[45] The group's parades—which drew thousands of cyclists every year from 1975 onwards, peaking at seven thousand in 1976—fall into this category.[46] So too do the MAB's mass die-ins at busy intersections and at the annual Montreal Auto Show. At the largest of these demonstrations, dozens of cyclists covered in ketchup and bandages halted traffic by sprawling across an intersection beside their bikes. Some were pronounced "dead" at the scene and others removed from the street on stretchers. Motorists halted by the die-in were told, "You are in the process of witnessing a hold-up. . . . We're not after money, we're after space." Onlookers were encouraged to lie down and participate in five minutes of silence, and some did. According to Silverman, the main goal of these dramas was to show observers an alternative to the current reality; for a few minutes cars would stop, silence would reign, and cyclists and pedestrians owned the road.[47]

Despite the intentionally humorous tone of these cyclo-dramas and their participants' attention-grabbing costumes and behaviour, they were nonetheless well planned, particularly when there was any danger of confrontation with motorists or the police. The die-ins, for example, were acted out according to detailed diagrams. Participants were divided into teams with specific roles, there was a minute-to-minute schedule, and after an attempt by a motorist to roll through an occupied intersection in 1976, provisions were made to protect the demonstrators with a "stalled" car.[48] Although dozens of MAB members were arrested during cyclo-dramas over the years, and in at least one case the arrestee complained of being roughed up, relations with the police generally remained cordial. In summer 1980, a busy moment for the MAB's metro access campaign, the police union refused for a time to arrest cyclists on the metro and expressed sympathy with their demands.[49]

The MAB's vélorutionary theatre drew on the tradition of street theatre protest developed in North America and Europe by the 1960s and 1970s counterculture. There are strong parallels, for example, to the theatrical "happenings" organized by the Diggers in New York and Toronto in 1967, in which so-called hippies blocked streets to cars, sometimes carrying cardboard replicas of traffic signs reading "Stop" and "No Parking."[50] More immediately, the MAB's tactics were inspired by its international networks; for example, one key early influence was a massive die-in organized in Australia in 1972, which the MAB heard about through contacts in Philadelphia. The MAB's strategic and highly publicized use of the die-in made it a model for other radical cycling organizations, and in 1977 one MAB member travelled to Amsterdam to share its strategies with groups there.[51] Since the MAB had been influenced from quite early on by Amsterdam's own Provos, there is an interesting symmetry in that visit. Combining elements of spectacle with a clear political message, the MAB's street theatre proved to be an effective means of reaching a broad audience.

(AHEAD) OF ITS TIME

The immediate reception of the MAB's theatrical demonstrations by the public was generally positive, despite the inevitable honks of frustrated drivers. The press gave ample coverage to the group's activities, partly because some individual journalists supported the MAB's demands and partly because cyclo-dramas made such good photo ops. For example, from May to December 1975, the MAB was featured in over forty newspaper articles in at least eight mainstream and alternative papers. The group also frequently generated debate in the form of letters to the editor. As part of a 1979 *Montreal Gazette* feature on metro access for cyclists, the paper printed sixteen letters on the issue, ranging in tone from supportive to hostile.[52]

Some columnists and letter writers criticized the MAB, calling the group's anti-car stance radical or unrealistic. For example, the *Sunday Express* scoffed at the idea that "being pro-bike requires that you be anti-car" and suggested that MAB members find something better to do with their time.[53] Overall, however, these critics were outnumbered by those who wrote in support of the MAB's demands—if not their tactics or larger critique of the automobile—citing the environmental and social benefits of supporting cycling in Montreal.[54]

While the MAB was successful at starting a public discussion on the place of cycling in the city, concrete government responses to its demands were slow to come. The group had some impressive successes in its early years, including the beginnings of a bike lane network and metro access for cyclists starting in 1983. But it would take decades of campaigning by the MAB and Vélo Québec (the new incarnation of La Fédération québécoise de cyclotourisme) for bike parking, public bicycles, separated lanes, and safe cycling links over the St. Lawrence to make it onto the agenda of Montreal's municipal government. Yet over the past forty-odd years nearly all of the demands made in the 1975 "Bicyclist's Manifesto" have been implemented, and in Montreal the idea that the bicycle is a viable form of urban transport has moved from the margins into the political mainstream. In a way, the MAB was ahead of its time. It was the first mass-membership

RECYCLING MONTREAL TRAFFIC

6.5 This cartoon, in which a surprised motorist is confronted by a gigantic MAB cyclist, ran in the *Gazette* on June 8, 1976. It expresses well the bewilderment of drivers confronted with bicycle advocates. Source: Terry Mosher/McCord Museum M-988.176.310.

group in Canada to focus its efforts on promoting city cycling, and in its deft combination of theatrical demonstrations and pragmatic lobbying it was as an inspiration for similar organizations across the country. Its calls for a society liberated from the automobile had less influence, perhaps, although echoes of that utopian vision can be seen in today's Critical Mass rides and the activities of groups like Montréal à Vélo. But, as this chapter has argued, the MAB was also very much *of* its time. Its foundation, ideas, and tactics were shaped by the specific historical context of 1970s Montreal: the vibrant activist networks and counterculture of Mile End and the Plateau, the authoritarian administration of Drapeau that became a rallying cry for oppositional movements, and, as historian Sean Mills has noted of the 1960s, the openness to international influences that characterized Montreal in the period.[55] In that context, cyclo-frustration, concern for the environment, and rejection of auto-centric culture all found a creative outlet in the MAB's unique brand of cycling advocacy. The MAB fought on behalf of Montreal cyclists and the environment for two decades, and after it lost momentum in the early 1990s many members continued to champion similar issues with a constellation of new organizations. MAB dynamo Claire Morissette would go on to found both Cyclo Nord-Sud, a not-for-profit that collects bicycles to send to the developing world, and the successful car-sharing service Communauto. Some in the 1970s, including the Civic Party administration, dismissed the MAB's vélorutionary demands as unrealistic or radical; in retrospect, they were the start of a conversation that has led to important changes in Montreal's urban environment.

NOTES

1   Thanks to Robert Silverman and the staff at Cyclo Nord-Sud for their enthusiastic co-operation. This chapter was researched and written with financial support from the Social Sciences and Humanities Research Council of Canada. It was published in a slightly different form as "'Vive la vélorution!': Le Monde à bicyclette et les origines du mouvement cycliste à Montréal, 1975–1980," *Bulletin d'histoire politique* 23, no. 2 (2015): 92–112.

2   Paul Rosen, "Up the Vélorution: Appropriating the Bicycle and the Politics of Technology," in *Appropriating Technology: Vernacular Science and Social Power*, ed. Ron Eglash (Minneapolis: University of Minnesota Press, 2004), 365–90; Zack Furness, *One Less Car: Bicycling and the Politics of Automobility*

(Philadelphia: Temple University Press, 2010).

3   See, for example, Richard Harmond, "Progress and Flight: An Interpretation of the American Cycle Craze of the 1890s" *Journal of Social History* 5, no. 2 (1971): 235–57; and G. B. Norcliffe, *The Ride to Modernity: The Bicycle in Canada, 1869–1900* (Toronto: University of Toronto Press, 2001).

4   Nina Dougherty and William Lawrence, *Bicycle Transportation* (Washington, DC: US Environmental Protection Agency, 1974), 5; Sharon Babaian, *The Most Benevolent Machine: A Historical Assessment of Cycles in Canada* (Ottawa: National Museum of Science and Technology, 1998), 86, 97.

5   "Les ventes ont doublé: Le public, surtout adulte, revient au

vieux cyclisme," *Dimanche-Matin*, 21 May 1972, quoted in Ivan Carel, "Les cyclistes: Du progrès moderne à la révolution écologiste," in *De la représentation à la manifestation: Groupes de pression et enjeux politiques au Québec, 19e et 20e siècles*, ed. Jérôme Boivin and Stéphane Savard (Quebec City: Septentrion, forthcoming).

6   Babaian, *The Most Benevolent Machine*, 86, 97.

7   Michael S. Goldstein, *The Health Movement: Promoting Fitness in America* (New York: Twayne, 1991).

8   Marc Raboy, interview with the author, 24 April 2012.

9   Vélo Québec, *L'État du vélo au Québec en 2000* (Montreal: Vélo Québec, 2001), 7; Vélo Québec, *L'État du vélo au Québec en 2010*, 15.

10   "Bicycle Touring Routes Proposed for City," *Montreal Gazette*, 31 May 1972; "La course extraordinaire jeudi: Le maire Drapeau n'était pas là pour recevoir sa bicyclette!" *Le Jour*, 27 May 1975.

11   James Stewart, "Jim Stewart's Montreal," *Montreal Star*, 10 April 1975.

12   See Andrée Fortin, *Le Rézo: Essai sur les coopératives d'alimentation saine au Québec* (Quebec City: IQRC, 1985), 39; and Sean Mills, *The Empire Within: Postcolonial Thought and Political Activism in Sixties Montreal* (Montreal: McGill-Queen's University Press, 2010), esp. 172.

13   Claire Helman, *The Milton-Park Affair: Canada's Largest Citizen-Developer Confrontation* (Montreal: Véhicule, 1987).

14   Raboy, interview; Robert Silverman, interview with the author, 10 April 2012.

15   Robert Silverman, "Bicyclist's Manifesto," May 1975, file "Manifeste cycliste," box 7, MAB Archives, Montreal (private holdings) (hereafter MAB).

16   Silverman, interview.

17   "Ils souhaitent un monde sur deux roues," *La Presse*, 2 June 1975; Raboy, interview.

18   Silverman, interview; MAB, "Nouveaux projets," *Bulletin*, no. 1 (1976).

19   Jean-Guy Vaillancourt, "Le mouvement écologiste québécois des années 80," in *Changer de société*, ed. Serge Proulx and Pierre Vallières (Montreal: Québec-Amérique, 1982), 141; Jane Barr, "The Origins and Emergence of Quebec's Environmental Movement: 1970–1985" (MA thesis, McGill University, 1995), 86.

20   See Fortin, *Le Rézo*, 38; and Timothy Thomas, *A City with a Difference: The Rise and Fall of the Montreal Citizens' Movement* (Montreal: Véhicule, 1997), 27.

21   Claire Morissette, "Automobile, je te hais!" *La Presse*, 9 March 1977.

22   "Let Us All Drive the Auto Show Out of Town," January 1976, file "Salon de la mort 76-01-09," box 7, MAB.

23   Silverman, "Bicyclist's Mani-
      festo."

24   Ibid.; Claire Morissette, "Les
      femmes et la vélorution," *Main-
      mise*, no. 76 (1978): 12.

25   Theodore Roszak, *The Making
      of a Counter Culture: Reflections
      on the Technocratic Society and
      Its Youthful Opposition* (Garden
      City, NY: Doubleday, 1969), 47,
      178–79; Paul Goodman and
      Percival Goodman, "Banning
      Cars from Manhattan," *Dissent*
      8, no. 3 (1961): 304–11; Andrew
      G. Kirk, *Counterculture Green:
      The Whole Earth Catalog and
      American Environmentalism*
      (Lawrence: University Press of
      Kansas, 2007), 28–31.

26   Christian Allègre, Michel Bélair,
      Michel Chevrier, and Georges
      Khal, *Le Répertoire québécois
      des outils planétaires* (Montreal:
      Mainmise, 1977).

27   Danielle Robinson, "Modernism
      at a Crossroad: The Spadina
      Expressway Controversy in To-
      ronto, Ontario, ca. 1960–1971,"
      Canadian Historical Review 92,
      no. 2 (2011): 295–322.

28   On anti-pollution activism in
      1970s Montreal, see Valérie Poi-
      rier, "'L'autoroute est-ouest, c'est
      pas le progrès!': Environnement
      et mobilisation citoyenne en
      opposition au projet d'autoroute
      est-ouest à Montréal en 1971,"
      Bulletin d'histoire politique 23,
      no. 2 (2015): 66–91.

29   Ivan Illich, *Energy and Equity*
      (London: Calder & Boyars,
      1974), 16.

30   The first use of the term "vélo-
      rution" seems to date to 1974,
      in France, and the presidential
      campaign of eccentric envi-
      ronmentalist Aguigui Mouna
      (André Dupont). See Jacques
      Danois, *Aguigui* (Bordeaux: Les
      Dossiers d'Aquitaine, 2007).

31   Silverman, "Bicyclist's Mani-
      festo."

32   Ibid.

33   Claire Morissette, *Deux roues,
      un avenir: Le vélo en ville*
      (Montreal: Écosociété, 2009),
      189.

34   "Compte rendu du congrès,"
      Bulletin, no. 5 (May 1977): 2–6

35   Morissette, *Deux roues, un
      avenir*, 191.

36   Silverman, "Bicyclist's Manifes-
      to."

37   Silverman, interview. For more
      on the Provos, see Richard
      Kempton, *The Provos: Amster-
      dam's Anarchist Revolt* (New
      York: Autonomedia, 2007).

38   "Les cyclistes cosmiques se réu-
      nissent," *Pour une ville nouvelle*
      3, no. 2 (May 1978): 10.

39   "Compte rendu du congrès";
      Silverman, interview.

40   Morissette, *Deux roues, un
      avenir*, 191.

41   "Bicycle Brigade Storms City
      Hall," *Montreal Gazette*, 6
      August 1975; Thomas, *A City
      with a Difference*, 54–7; Richard
      Wagman, "Lobbying ou mobili-
      sation?" *Pour une ville nouvelle*,
      no. 2 (December 1977): 6.

42 "Le Monde à bicyclette aménage sa propre piste," *Le Devoir*, 24 July 1978.

43 "Cyclist in Cell Loved View of Bikers' Path," *Montreal Gazette*, 10 October 1981.

44 "Even a Hippo Can't Get Montreal Cyclists on Metro," *McGill Daily*, 15 September 1980; MAB, "Opération Moïse," file "Moïse 81-04-20," box 6, MAB.

45 Morissette, *Deux roues, un avenir*, 205.

46 "7,000 partisans du vélo ont pédalé dans Montréal," *Le Jour*, 7 June 1976.

47 "Mourir 5 minutes avec le monde à bicyclette," *La Presse*, 12 October 1976; MAB, "Dear Motorist," 12 October 1976, file "Crime quotidien II 76-10-12," box 7, MAB; Silverman, interview.

48 MAB, "Déroulement: Journée mondiale contre l'auto," 11 October 1977, file "Journée mondiale contre l'auto 77-10-11," box 7, MAB; Silverman, interview.

49 MAB member (identity confidential), "Description des faits du 30 août 1978," file "Métro accès procès," box 7, MAB; "La police refuse d'arrêter des cyclistes dans le métro," *Le Journal de Montréal*, 19 June 1980.

50 Michael William Doyle, "Staging the Revolution: Guerilla Theater as a Countercultural Practice, 1965–68," in *Imagine Nation: The American Counterculture of the 1960s and '70s*, ed. Peter Braunstein and Michael William Doyle (New York: Routledge, 2002), 86; "Hippies Chant Love to 'Polite' Policemen," *Toronto Star*, 22 August 1967.

51 Silverman, interview; Morissette, *Deux roues, un avenir*, 188.

52 "Let Cyclists onto Metro," *Montreal Gazette*, 24 August 1979.

53 See, for example, "Hook, Line and Sinclair," *Sunday Express*, 21 March 1976; or "Le métro, pour les piétons!" *Journal de Montréal*, 28 November 1978.

54 Louis Fournier, of *Le Devoir* and *Le Jour*, was a particularly staunch ally of the group. See "Le Monde à bicyclette demande: 'Etiez-vous obligé de prendre votre auto aujourd'hui?'" *Le Devoir*, 25 September 1975.

55 Mills, *The Empire Within*, esp. 19–84.

SECTION 2:

PEOPLE, NATURE, ACTIVITIES

An Ark for the Future: Science, Technology, and the Canadian Back-to-the-Land Movement of the 1970s

Henry Trim

The future arrived at Spry Point, a secluded area on the eastern end of Prince Edward Island, in September 1976. It came in the form of a "Space Age Ark."[1] A large structure designed to use renewable energy and to provide food for its inhabitants, the Ark bioshelter responded to Canadian concerns about energy use and out-of-control development. This unique building became national news as Premier Alexander Campbell and Prime Minister Pierre Elliott Trudeau flew in by helicopter to attend its opening. Leading members of the "appropriate technology"[2] movement, a goodly number of hippies, and a few somewhat incredulous islanders also attended the opening ceremony, celebrating late into the night.[3] Addressing this diverse group Trudeau proclaimed that the Ark bioshelter would be an example to those who wished "to live lightly on the earth," and Dr. John Todd, the Ark's principal designer, stated that its "small is beautiful" approach

7.1 Opening the Ark on September 21, 1976. Left to right: Premier Alexander Campbell, John Todd, Nancy Jack Todd, Prime Minister Pierre Elliott Trudeau. Source: *An ARK for Prince Edward Island: A Report to the Government of Canada from New Alchemy Institute*, Little Pond, RR4, Souris PEI (902) Cardigan 181, 30 December 1976.

to development would show Canadians how to live within nature's limits.[4]

The Ark, with its space-age technology, scientist designers, and government funding, does not conform to usual expectations of a countercultural project; in fact, it directly challenges the dominant understandings of the counterculture. In 1969, Theodore Roszak—whose work defined initial analysis of the counterculture—described it as a utopian youth movement that opposed Western rationality, particularly science and technology, and sought spiritual enlightenment.[5] Recently, however, historians—led by Fred Turner and Andrew Kirk—have questioned whether romantic youth suspicious of science and technology and out to harass or escape authority really defined the counterculture. The work of these historians has pointed to a pragmatic side of the counterculture that embraced science and technology and involved scientists, engineers, and government as well as alienated youth.[6]

The Ark challenges those who have applied Roszak's views to the Canadian counterculture.[7] The analysis of this experiment provides a more complete understanding of how some groups employed technological solutions when dealing with environmental challenges. Designed and built by countercultural scientists from the New Alchemy Institute in Cape Cod, Massachusetts, the novel structure highlights the importance of scientific knowledge and technological innovation to the counterculture. For the New Alchemists, this focus on technology proved useful as it expanded the group's influence. In particular, it played an important role in the provincial and federal governments' decision to provide hundreds of thousands of dollars in funding for the Ark. This support also suggests that appropriate technology advocate E. F. Schumacher's small-is-beautiful approach to development in the 1970s enjoyed a substantial degree of popularity among Canadians.[8]

The New Alchemists' technophilia also highlights problems inherent in the counterculture's embrace of technology. Langdon Winner, for instance, has argued that this focus on technology at times became myopic and led some to neglect other avenues for social change.[9] Among the New Alchemists, it inspired technological optimism—specifically, the belief that new technology had the potential to transform Canada into the participatory and sustainable society they desired. This technological optimism resulted in a substantial discontinuity between the New Alchemists' rhetoric and their results, mirroring the broader movement's difficulty in achieving its ambitious goal of setting an example for a better society by going back to the land. Burdened with these high expectations and hampered by technical problems, the Ark would malfunction and disappoint its supporters rather than start the hoped-for transformation of PEI.

Despite its failure, the Ark, and the movement it represented, had a significant impact on Canadian society. Most notably, it helped to introduce Canadians to renewable energy and organic foods as well as pioneering green architecture, aquaponics, and sustainable farming. As a primogenitor of these developments, and as an example of diversity within the counterculture and the support some of its ideas enjoyed in Canada, the Ark stands as an important piece of Canada's countercultural history. While downplaying the countercultural ethos of the Ark, Alan MacEachern's excellent history of the Institute of Man and Resources and its experiments with renewable energy and alternative development on PEI focuses extensively on the experiment.[10] Expanding upon MacEachern's account, this paper places greater emphasis on the New Alchemists and their technological and countercultural vision, to better connect events on PEI with the wider youth movement and to highlight the importance of technological optimism in the Canadian counterculture.

THE VISION

In 1969, Canadian ethologist John Todd and American marine biologist William (Bill) McLarney founded an institute dedicated to providing scientific assistance to the back-to-the-land movement. Motivated both by environmental concerns and by their first-hand experience of the difficulties of going back to the land at a short-lived commune in rural California, the two established the oddly named New Alchemy Institute (NAI).[11] The institute's charter states that the group planned to "engage in scientific research in the public interest on ecologically and behaviourally planned agriculture systems and rural land based communities."[12] As an organization, the NAI united scientists, anti-war protesters, and commune-dwellers to assist in the counterculture's search for social justice and the environmental movement's attempts to protect and restore the environment.[13] To carry out their self-assigned mission, the New Alchemists set up their institute on Cape Cod in 1971, near the Woods Hole Oceanographic Institute, where Todd and McLarney had worked before dedicating themselves full-time to the NAI.[14]

In organizing their institute, the New Alchemists drew heavily upon practices of the counterculture; indeed, they structured it along the lines of a countercultural commune.[15] Rejecting the hierarchical organization that pervaded the scientific institutions Todd and McLarney had left behind, the NAI adopted an individualistic and egalitarian organizational structure based on the participatory models of the New Left.[16] Every member of the NAI was officially equal and free to pursue what interested him or her. This philosophy attracted members of the counterculture and young scholars who shared Todd's and McLarney's environmental and social concerns and their optimistic view of science and technology. Their fusion of technology and counterculture also enjoyed a good deal of popularity in the 1970s. In fact, the New Alchemists were part of a subsection of the counterculture centred on the *Whole Earth Catalog* that employed a distinctive approach to social and environmental problems.

These "countercultural environmentalists," as Kirk has called them, were enamoured by the possibility of constructing whole systems incorporating man, machine, and nature within a single sustainable structure.[17] In their view, such systems had the potential both to protect the environment and to realize the counterculture's goal of a participatory society. Founded upon arguments popularized by R. Buckminster Fuller and Schumacher, among others, their approach argued that technology had a deep impact on both the environment and society.[18] For them, technology mediated human interactions with nature and formed the foundation of all social structures. Small-scale, easily intelligible technologies, for instance, were viewed as inherently democratic—a form of "technology with a human face"—since they encouraged decentralization and could be understood by everyone.[19] Thus, for these members of the counterculture, technological change played a central role in any social or environmental transformation, since the adoption of new technologies could alter social structures and human relationships with the environment.[20]

Ecology also had a central place in this "countercultural environmentalism." The designs of the New Alchemists and other countercultural environmentalists drew heavily on the systems ecology of Howard T. Odum and Eugene Odum. The Odum brothers employed cybernetics to merge humans, technology, and nature into a single

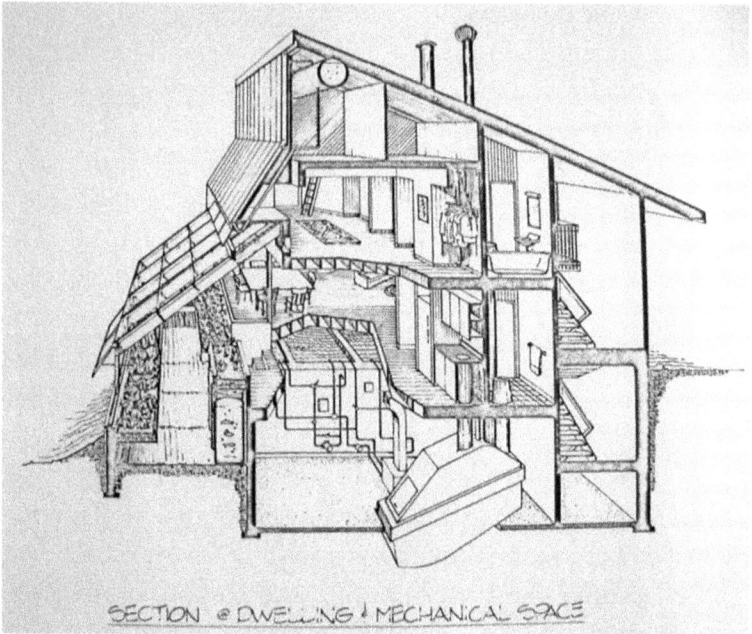

SECTION @ DWELLING & MECHANICAL SPACE

7.2 Interior of the PEI Ark. Note the spacious rooms and the composting system in the basement connected to the kitchen and bathrooms. Source: *An ARK for Prince Edward Island.*

feedback system.[21] This research suggested the possibility of designing a system to be almost completely self-sufficient, thus sustainable and well suited to a decentralized society. NASA, in fact, attempted something along these lines as it worked with ecologists to design self-contained ecosystems capable of supporting astronauts on lengthy missions.[22] This added further inspiration to countercultural environmentalists' desire for self-sufficient systems. Embracing NASA's research on space capsules as both a design approach and a metaphor for understanding the global ecosystem, Fuller, Stewart Brand, and the *Whole Earth Catalog* helped to popularize the "spaceship earth" concept in the 1960s and 1970s.[23]

The Ark project brought together countercultural environmentalists' ideas about ecology, technology, and society and the back-to-the-land movement's desire to live sustainably on the land in a way that few other projects did.[24] Building on these ideas, the New Alchemists designed their Ark to achieve the long-standing goal of countercultural environmentalists: to create a technology that allowed back-to-the-landers to combine "agriculture, aquaculture, and power generation . . . to enable [them] to satisfy [their] needs without destroying the resources which provide them."[25] This made the PEI Ark an odd sort of "spaceship to the future," as one journalist dubbed it, since it promised to transport Canadians to a high-tech decentralized society, powered by renewable energy and scientifically managed to maintain the earth's ecological balance.[26] In short, the Ark was to be an elegant method for using technology to remake Canadian society and protect the environment.

GOVERNMENT INTEREST

The 1970s energy crisis provides the essential backdrop for Canadian government interest in the New Alchemists' experiments. In a decade largely defined by the rise of environmentalism and the neo-Malthusian "limits to growth" thesis and oil shocks, advocates of small-is-beautiful ideas were able to force their way into the discussion of Canada's future.[27] Even the Science Council of Canada, an elite technocratic advisory body founded in 1966, became a strong advocate of alternative energy; in 1973, it devised the "conserver society" and championed sustainable development for the rest of the decade.[28] As a result, the question of whether Canada would continue down the "hard technology" and "high energy" path it had followed since the end of World War II or shift toward the "soft technology" small-scale development strategy advocated by countercultural environmentalists became a point of national discussion.[29]

While the energy crisis explains interest in the "conserver society" and renewable energy, three further reasons led federal and provincial governments to fund the New Alchemists specifically. First, Canadian media explained the group's work in very positive terms. Journalist Barry Conn Hughes, for example, told Canadians that the New Alchemists had devised a system "which could feed itself" without relying on oil.[30] Second, Todd's salesmanship and his ability to fascinate an audience as he expounded upon the bright future of renewable energy and small-is-beautiful development brought him to the attention of Canadian governments and helped him to make contacts in Ottawa and Charlottetown.[31] Finally, and most importantly, Todd's scientific credentials and the New Alchemists' innovative experiments with self-sufficiency carried weight with government officials. For instance, visiting the NAI and talking with Todd convinced Robert Durie, the director of the Advanced Concepts Centre at Environment Canada, that the New Alchemists' work could help Canada deal with its energy needs, and he avidly supported funding the group.[32]

Electoral politics assisted the group as well. As historians Wayne MacKinnon and Alan MacEachern suggest, funding groups such as the New Alchemists allowed Prime Minister Trudeau and Premier Campbell to win support among both environmentalists and members of the counterculture with little risk.[33] In economically depressed Atlantic Canada, funding the New Alchemists' project could also contribute to the Trudeau government's efforts to spark regional economic development.[34] In short, as federal and provincial governments searched for new approaches to energy use and economic development, the countercultural scientists of the NAI seemed to offer credible solutions and possible political gains at little cost.

The New Alchemists' promises had the greatest appeal in PEI. Completely reliant on imported oil, the province faced a bleak future as it seemed that oil prices would climb indefinitely. Oil had jumped from about three dollars a barrel in 1973 to eight dollars a barrel by 1975, making the prospect of further price increases very likely.[35] Concerned by the future of his province, and uneasy about the sustainability of the high-energy society in general, Liberal Premier Campbell and his closest advisor, Andy Wells, began to examine alternative paths of development.[36] The decentralized and small-scale approach advocated by Schumacher became Wells's and Campbell's preferred approach to island development.[37] This small-is-beautiful development model emphasized renewable energy, local resources, and simple, labour-intensive technologies.[38] Eager to begin experimenting with strategies for alternative development, Campbell began calling for greater support for renewable energy at the federal level.[39] In PEI, he founded the Institute of Man and Resources (IMR), a research institution meant to spearhead the development of renewable energy and locally appropriate industry. The IMR quickly launched Energy Days, a four-day investigation and discussion of PEI's energy future and Canada's energy options held in the summer of 1976.[40] These efforts gained results as Ottawa agreed to fund the development of renewable energy on the island early in 1977.[41] Although largely forgotten outside the province, the Canada-PEI Agreement on Renewable Energy Development briefly made PEI a leading centre of renewable energy research and development within Canada.[42]

During this push to investigate and experiment with alternatives, Campbell and Wells invited the New Alchemists to set up an institute in PEI. Initial funding came from Environment Canada and Urban Affairs Canada through Canada's UN Habitat 1976 project.[43] The Ark's proposed self-contained food-producing systems fit well with Habitat's focus on sustainable urban development, giving the two ministries and the province of PEI an opportunity to share the costs of a project that was of interest to all.[44] With funding secured, the New Alchemists' Ark quickly became a central, or at least the most publicized, component of PEI's efforts to apply small-is-beautiful thinking to the challenges of the energy crisis.[45]

Todd pragmatically seized the opportunity to work with the provincial and federal governments and gain access to the funds necessary to make his ideas a reality. Optimistic about the project, Todd promised that the New Alchemists would provide Prince Edward Islanders with a low-cost ecologically derived structure "designed to sustain their food, shelter, and power needs."[46] With such statements about the Ark, Todd downplayed its experimental nature and portrayed it as a straightforward solution to the problems facing islanders. The financing the Ark received illustrates the important role the Canadian federal and provincial governments occasionally played in the Canadian counterculture and back-to-the-land movement.

However, this support exposed the movement to public scrutiny. If groups such as the New Alchemists could not achieve their stated goals, they risked dismissal as failures and squanderers of public funds.

PUBLIC VS. PRIVATE

In 1975, even before construction of the Ark, the New Alchemists began to have problems. The first was a clash of cultures between conventional islanders and the countercultural New Alchemists.[47] Attempting to assuage local fears, Todd and his colleagues held town hall meetings to explain the New Alchemists' work and their desire to assist the small-scale farming communities on PEI with their research at the forthcoming Ark. Most islanders, however, remained unconvinced that the New Alchemists' work would be of any use to them and were suspicious of the primarily American group that had landed in their midst. Such tension between locals and back-to-the-land groups did not occur everywhere, but when it did, it could easily derail countercultural attempts to construct a better future.[48]

Problems with the public continued as the Ark gradually took shape over the summer of 1976, stemming primarily from the New Alchemists' complete surprise at the nation-wide interest in their work and their inability to benefit from this attention.[49] Curious locals, tourists, and travelling hippies visited the site, interested in the odd futuristic structure and the reasons for its construction. This level of interest in a solar- and wind-powered structure may seem odd today, but it was the first building of its kind in Prince Edward Island and one of the very first "green" buildings in Canada. Unfortunately, rather than engaging visitors, the New Alchemists worried that they would not complete construction on schedule; trying to stretch their resources as far as possible, they responded by putting up signs instructing visitors not to talk to the carpenters.[50]

Canadians' curiosity grew even greater after the Ark opened, as the grand opening, attended by Prime Minister Trudeau, generated national media coverage. Stories about the Ark and its promise of a

7.3 The PEI Ark viewed from the south. The greenhouse is in the right foreground and the living area in the left background. Source: *An ARK for Prince Edward Island*.

sustainable future quickly appeared in national and regional magazines such as *Chatelaine, Harrowsmith*, and the *Atlantic Advocate*.[51] In response, thousands visited the Ark every year, and it became something of a pilgrimage site among the back-to-the-land movement. Unable to cope with this number of visitors and still carry out their research, the New Alchemists living and working in the Ark began barring the gate in an attempt to restrict visitation to Wednesday afternoons and Sundays. This seemed a reasonable decision from their perspective, since they saw the Ark as a private research facility and believed the research they conducted there would lead to a broader social and environmental transformation of the island. By dismissing visitors to focus on developing their technology, the New Alchemists

7.4 The PEI Ark's greenhouse. The large plastic cylinders are solar algae ponds for raising fish and moderating the greenhouse temperature. Source: *An ARK for Prince Edward Island.*

passed up a rare opportunity to build public support for their work and their goals of social transformation.

Unsurprisingly, this disregard frustrated those interested in the Ark. The New Alchemists' stance particularly rankled because they had received government funding to demonstrate self-sufficient methods of living. Curious to see what their tax dollars had paid for, visitors viewed the Ark as a fully public facility, and they responded to the New Alchemists' limitation of visitation by demanding entry to the Ark. When New Alchemists tried to turn away *Harrowsmith* reporter David Lees, even this ardent supporter of the back-to-the-land movement argued that his taxes granted him a right to enter the building.[52]

Confusion over whether the Ark was a private research facility or a public site stemmed partly from the differing goals of the New Alchemists and Canada's federal and provincial governments. The New Alchemists, seeing their mission as one of research and development, had little interest in using the Ark as a demonstration project for renewable energy and sustainable living; however, demonstration played a key role in their federal sponsors' hopes that the Ark would help to educate Canadians about these issues.[53] An emphasis on demonstration pervaded the press coverage of the Ark as well. Constance Mungall's article in *Chatelaine*, for example, depicted the Ark as a domestic space that represented a new type of home for Canadians.[54] Confusion over its goals also led to internal conflicts between New Alchemists and their provincial managers in the IMR.[55]

This conflict came to a head in late 1977. Frustrated by the never-ending stream of visitors into their home and workspace, David Bergmark and Nancy Willis—the New Alchemists who had been living in the Ark to assess its utility as a house—moved out. With their departure, the hope that the Ark would showcase sustainable family living ended. The couple's departure undermined a fundamental reason for the building the Ark and damaged the province's faith in the countercultural group's ability to run the structure successfully.

TECHNOLOGIES

While the New Alchemists faced difficulties with the public and confusion about their mission, considerable problems also beset the technologies they attempted to develop. In fact, despite Todd's downplaying of the Ark's experimental nature and his assurances that earlier prototypes had perfected the technology, the structure never functioned properly. Designed using a "spaceship approach," the Ark recreated the intricate systems of a stable ecosystem.[56] The New Alchemists believed their design would allow them to mimic the self-sufficiency of natural ecosystems while providing food and shelter for its inhabitants.[57] The New Alchemists also hoped the design would demonstrate that Canadians could live within the limits of a closed ecosystem. Drawing inspiration from Schumacher, the New Alchemists saw the Ark as an "adaptive structure" capable of transforming Canada into a sustainable and participatory society.[58]

At the core of the Ark lay a solar greenhouse irrigated by an interconnected series of fishponds. The focus of years of research at the NAI, the solar greenhouse's combined aquaculture and agriculture system worked very well.[59] In an elegant system of the New Alchemists' own devising, the aquaculture ponds played a central role in a managed nutrient cycle: pond water fertilized plants while plants and bacteria filtered the pond water for fish. To ensure that the system continuously recycled nutrients as effectively as possible, the New Alchemists managed their greenhouse system entirely without pesticides or synthetic fertilizers.[60] Although not quite "the world that feeds itself" that some claimed, the New Alchemists' integrated system recorded substantial levels of fish and vegetable production with minimal inputs.[61]

Unfortunately, not everything worked quite so well for the New Alchemists. Embarrassingly for the group, some of the technological components, which they claimed to have thoroughly tested, failed to function. The integrated systems of the Ark exacerbated these problems as the building's complex internal feedback systems conflicted with each other, further damaging its operation. One of these malfunctioning subsystems was the solar heating and air circulating system. In an effort to understand and manage these systems, the New Alchemists installed a state-of-the-art computer in 1976.[62] At the time, computers were expensive and delicate pieces of hardware largely unknown outside science labs. The installation of one pushed the cost of the Ark into the hundreds of thousands of dollars, far beyond the means of most Canadians. Although necessary, installing the computer weakened the credibility of the New Alchemists, since they had always claimed that their technology could easily be adapted for broad adoption. Indeed, this had been the fundamental point of the Ark's "adaptive" design and the method through which they hoped to change Canadian society.[63] The structure's complexity also meant that managing it required a considerable degree of training and knowledge, which further undermined the New Alchemists' stated desire to produce easily intelligible technologies that everyone could use.

Even more damaging to both the Ark and the New Alchemists' reputation was the failure of the Ark's wind turbines. Meant to demonstrate the Ark's self-sufficiency and launch the island toward a wind energy program, the turbines were central to the New Alchemists' research as well as to the broader small-is-beautiful development program for PEI.[64] Wells, for instance, specifically highlighted the New Alchemists' turbines when discussing the Ark with islanders, arguing that the turbines had the potential to start a wind industry on PEI.[65] Completely designed by the New Alchemists, the turbines employed a novel system using hydraulics to control the blades and generate electricity.[66] In an effort to construct the Ark as quickly as possible and meet the September opening deadline set by Prime Minister Trudeau, the New Alchemists deployed their experimental turbine without extensive testing.[67] Overwhelmed by PEI's high winds, the turbine's untested hydraulics soon seized up, forcing the Ark ignobly to draw electricity from PEI's grid.[68]

As frustrating as these failures were, they might not have damaged the project had Todd not optimistically assured locals of success nor consistently de-emphasized the experimental nature of the Ark. With expectations further raised by the initially laudatory media attention, Prince Edward Islanders expected great things from the Ark. Instead, the Ark experienced a cascade of problems, as experiments often do. For islanders who had been all but promised success, and even a new industry, the malfunction of the New Alchemists' turbines and the Ark's reliance on PEI's grid defined the project as a failure.

A degree of prejudice among some islanders also sped their dismissal of the Ark. As Wells later recalled in an interview, some islanders harboured deep suspicions about the small-is-beautiful approach and happily criticized the Ark at every opportunity.[69] Jim MacNeil, the editor of the *Eastern Graphic* and the unofficial leader of skeptical islanders, had been critical of the Ark from the very beginning. His editorial on the opening of the Ark focused on the fuel wasted by flying Trudeau to the opening ceremony.[70] With the very public airing of the Ark's growing problems, the structure's $354,000 price tag and the tens of thousands of dollars spent on annual costs began to rankle.[71] By 1978, the New Alchemists faced suggestions in the local press that they had wasted tax dollars and even swindled the Canadian government.[72] This bad press, compounded by the New Alchemists' fumbling public relations and narrow focus on research, did little to change islanders' view of the Ark, and the project gradually turned from an asset to a political liability that threatened the provincial government's hopes for a small-is-beautiful approach to development.[73]

Disenchanted by the partial successes of the New Alchemists, Environment Canada distanced itself from the Ark and began to withdraw funding from the program in 1978.[74] Faced with financial difficulties and increasingly strict oversight from the IMR, along with mounting technical and public relations problems, the New Alchemists decided to abandon their work at the PEI Ark and concentrate instead on their institute at Cape Cod. In February 1978, they handed over the Ark to the IMR.[75] This marked the end of the New Alchemists' time on PEI and the end for one of Canada's best-known examples of the small-is-beautiful approach to development.

The Ark itself continued to function for another two years, under the direction of Ken MacKay, a biologist the IMR hired to take over its supervision. Research into organic gardening, non-chemical pest control, and small-scale aquaculture as well as tours and demonstration projects continued with relatively little interruption. However, with government money for novel solutions to the energy crisis drying up and the election of a new and unfriendly provincial government in 1980, the Ark entered a period of financial limbo; it closed permanently in 1981.[76] After sitting vacant for nearly two decades, the Ark was demolished in 2000 to make room for the Inn at Spry Point.[77]

CONCLUSION

Despite its ultimate failure, the PEI Ark had a considerable impact during the 1970s and left a significant legacy. In the turbulent 1970s, the Ark helped to popularize the concepts of renewable energy and sustainability among Canadians. Environmental groups showed particular interest in the New Alchemists' work, as it seemed to offer a reasonable solution to pressing concerns about natural limits and energy conservation. In fact, Pollution Probe of Toronto put many of the New Alchemists' ideas to work in their ecology house project, which included both energy conservation and a solar greenhouse.[78] Research conducted at the Ark also led directly to improvements in natural pest control, a significant development for organic gardeners and farmers.[79] The New Alchemists even briefly enjoyed a reputation as national experts on renewable energy and sustainability. When the Department of Energy, Mines, and Resources began to consider a federal program for renewable energy in 1977 it sought out Todd to serve on the National Advisory Committee on Conservation and Renewable Energy.[80] In short, the Ark became a widely known, if flawed, example of the small-is-beautiful approach in Canada.[81]

The New Alchemists themselves learned a great deal from the Ark and its problems. Immediately following their departure from PEI, the group began to distance themselves from their overly optimistic goal of self-sufficiency. Instead, they concentrated on what had worked in the Ark—the solar greenhouse with its combined aquacultural and agricultural systems—and used this "living technology" to design ecologically sustainable urban farming and waste management systems.[82] Marking the completion of this transition in 1981, the New Alchemists published a special issue of their journal focused on urban agriculture and solar design.[83] In fact, the feedback loops designed into the Ark's greenhouse directly prefigured the emergence of "aquaponics," a highly efficient approach to greenhouse agriculture that combines aquaculture and hydroponics.[84] Building on their work on the PEI Ark, John Todd and Nancy Jack Todd went on to play a significant role in the American "green architecture" movement.[85]

Beyond their direct and indirect legacies, the New Alchemists and their Ark encourage historians of the counterculture and the back-to-the-land movement to recognize the important role science and technology played within both. Together, science, technology, and the counterculture shaped how countercultural environmentalists, such as the New Alchemists, attempted to change Canadian society for the better. Besides revealing the technological side of the counterculture, the Ark's failure warns against technological optimism and underlines the importance of local politics and social engagement to any attempt to effect social change. The example of the New Alchemists also reveals the maturity and pragmatism of the Canadian counterculture and back-to-the-land movement. Ready to work with provincial and federal governments to further their goals, the New Alchemists were not the young, romantic, anti-authority hippies who too often overshadow images of the Canadian counterculture. The history of the Ark also illustrates the different levels of government recognition of the counterculture and the concerns of governments as they directly supported some of the efforts to change Canadian society and protect the environment. This willingness, particularly on the part of Premier Campbell, demonstrates the influence of small-is-beautiful ideas during the crisis-wracked 1970s as one of Canada's provinces made such ideas a central part of its energy and development policy.

NOTES

1 Constance Mungall, "Space Age Ark: A Brave New Home," *Chatelaine*, November 1977, 52–53, 103–9.

2 The appropriate technology movement combined an antipathy towards large technological systems, such as nuclear power, with a belief that small technologies, such as solar panels, could both protect the environment and rejuvenate society. Its most prominent leader was E. F. Schumacher, while Stewart Brand's *Whole Earth Catalog* provided its main voice. For a general discussion of the appropriate technology movement, see Jordan Kleiman, "The Appropriate Technology Movement in American Political Culture" (PhD diss., University of Rochester, 2000).

3 Nancy Jack Todd, "The Opening of the Arks," *Journal of the New Alchemists*, no. 4 (1977): 10–17.

4 Stewart Brand, "From Urgencies to Essentials," *CoEvolution Quarterly*, no. 13 (Spring 1977): 166–67; N. Todd, "The Opening of the Arks,"16; Alan MacEachern, *The Institute of Man and Resources: An Environmental Fable* (Charlottetown, PEI:

Island Studies, 2003), 35; Andrew Kirk, *Counterculture Green: The Whole Earth Catalog and American Environmentalism* (Lawrence: University of Kansas Press, 2007), 147.

5   Theodore Roszak, *The Making of a Counterculture: Reflections on the Technocratic Society and Its Youthful Opposition* (New York: Anchor, 1969).

6   Kirk, *Counterculture Green*, 12; Fred Turner, "Where the Counterculture Met the New Economy: The WELL and Origins of Virtual Community," *Technology and Culture* 46, no. 3 (2005): 485–512; Fred Turner, *From Counterculture to Cyberculture: Stewart Brand, the Whole Earth Network, and the Rise of Digital Utopianism* (Chicago: University of Chicago Press, 2006); Peter Braunstein and Michael William Doyle, "Introduction: Historicizing the American Counterculture of the 1960s and '70s," in *Imagine Nation: The American Counterculture of the 1960s and '70s*, ed. Peter Braunstein and Michael William Doyle (New York: Routledge, 2002), 1–14.

7   Roszak's misinterpretation of the counterculture has influenced some important Canadian analyses. See Brian Palmer, *Canada's 1960s: The Ironies of Identity in a Rebellious Era* (Toronto: University of Toronto Press, 2009), 205–9; and Doug Owram, *Born at the Right Time: A History of the Baby Boom Generation* (Toronto: University of Toronto Press, 1996), 185–87.

8   E. F. Schumacher, *Small Is Beautiful: A Study of Economics as if People Mattered* (London: Blond & Brigs, 1973). Schumacher's ideas were often combined with energy analyst Amory Lovins's "soft path" to form a relatively well-developed alternative approach to economic and energy development. Amory Lovins, *Soft Energy Paths: Toward a Durable Peace* (San Francisco: Friends of the Earth, 1977). The "small is beautiful" approach adhered to a few fundamental concepts that both the New Alchemists and PEI's provincial government regarded as important. The most fundamental principle was that a network of small, simple technological systems was both more efficient and more durable than a large, complex, centralized technology. Networks of simple technologies, according to the theory, also had the added advantages of encouraging extensive local development and increasing local political power as well as decreasing environmental impacts. A good illustration of this approach can be seen in the comparison of a nuclear power plant with a network of user-operated photovoltaic arrays and solar heaters. Both systems provide electricity and heat but use a very different organizational approach, with divergent levels of technological complexity, and have (theoretically) contrasting political consequences.

9   Langdon Winner, *The Whale and the Reactor: The Search for Limits in an Age of High Technology*

(Chicago: University of Chicago Press, 1986), 58–84.

10   See MacEachern, *Institute of Man and Resources.* I would also like to thank Dr. MacEachern for his assistance with the research for this paper.

11   John Todd, "Introduction," *New Alchemy Institute Bulletin,* no. 1 (Fall 1970): 1; John Todd, "The New Alchemists," in *Design Outlaws on the Ecological Frontier,* ed. Chris Zelov and Phil Cousineau (Easton, PA: Knossus, 1997), 175.

12   New Alchemy Institute, *Articles of Incorporation,* San Diego, CA, 1970, box 1, folder 11, New Alchemy Institute Records, MS 254, Special Collections, Iowa State University Library, Ames, Iowa.

13   Nancy Jack Todd, "Readers Research Program," *New Alchemy Newsletter,* no. 1 (Spring 1972): 11; John Todd, "A Modest Proposal," *New Alchemy Institute Bulletin,* no. 2 (Spring 1971): 1–12.

14   Nancy Jack Todd, "New Alchemy Institute—East: Cape Cod," *New Alchemy Newsletter,* no. 1 (Spring 1972): 4.

15   Timothy Miller, *The 60s Communes: Hippies and Beyond* (Syracuse: Syracuse University Press, 1999), 139–42, 149–50.

16   John Todd, "Realities from Ideas, Dreams and a Small New Alchemy Community," *New Alchemy Newsletter,* no. 2 (Fall 1972): 5–6. For a discussion of Cold War universities, see Rebecca Lowen, *Creating the Cold War University:*

The Transformation of Stanford (Berkeley: University of California Press, 1997). For a discussion of the participatory models of the New Left, see Francesca Polletta, *Freedom Is an Endless Meeting: Democracy in American Social Movements* (Chicago: University of Chicago Press, 2002), 122–30; and Myrna Kostash, *Long Way from Home: The Story of the Sixties Generation in Canada* (Toronto: J. Lorimer, 1980), 7–8, 133–34.

17   Kirk, *Counterculture Green,* 17.

18   Schumacher, *Small Is Beautiful,* 21, 167; R. Buckminster Fuller, *Operating Manual for Spaceship Earth* (Carbondale: Southern Illinois University Press, 1969). See also Lewis Mumford, "Authoritarian and Democratic Technics," *Technology and Culture* 5, no. 1 (1964): 1–8; Jacques Ellul, *Technological Society,* trans. John Wilkinson (New York: Vintage, 1964); Barry Commoner, *The Closing Circle: Nature, Man, and Technology* (New York: Knopf, 1971); and Herbert Marcuse, *One Dimensional Man: Studies in the Ideology of Advanced Industrial Society* (Boston: Beacon, 1964).

19   Schumacher, *Small Is Beautiful,* 138.

20   There is some debate over the degree to which the appropriate technology movement actually held deterministic views, but it is beyond the scope of this paper. See Frank Laird, "Constructing the Future: Advocating Energy Technologies in the Cold War," *Technology and Culture* 44, no. 1

(2003): 27–49; Langdon Winner, "The Political Philosophy of Alternative Technology," *Technology in Society* 1, no. 1 (1979): 75–86.

21 Sharon Kingsland, *Evolution of American Ecology, 1890–2000* (Baltimore: Johns Hopkins University Press, 2005), 190–92; Howard Odum, *Environment, Power, and Society* (New York: Wiley-Interscience, 1970); Eugene Odum, *Fundamentals of Ecology*, 3rd ed. (Philadelphia: Saunders, 1971).

22 Kirk, *Counterculture Green*, 170–81; Allen Brown, "Regenerative Systems," in *Human Ecology in Space Flight*, ed. Doris Calloway (New York: New York Academy of Sciences, 1963), 82–119, esp. Eugene Odum's comments on 85–87.

23 Peder Anker, "Ecological Colonization of Space," *Environmental History* 10, no. 2 (2005): 239–68. Fuller's compelling vision inspired the architecture of many other designs similar to the New Alchemists' Ark, such as the recent Eden Project in Britain. See Nicholas Grimshaw, *The Architecture of Eden* (London: Eden Project, Transworth, 2003).

24 Robert Angevine, Earle Barnhart, and John Todd, "New Alchemy's Ark: A Proposed Solar Heated and Wind Powered Greenhouse and Aquaculture Complex Adapted to Northern Climates," *Journal of the New Alchemists*, no. 2 (1974): 35–44.

25 J. Todd, "Introduction," 1–2; Nancy Jack Todd, "Bioshelters and Their Implications for Lifestyle," *Habitat International* 2, nos. 1–2 (1977): 87–100.

26 Walter Stewart, "The Ark: PEI's Spaceship to the Future," *Atlantic Advocate*, September 1976, 42–44.

27 Barry Conn Hughes, "The World that Feeds Itself," *Canadian Magazine*, 9 February 1974, 2–6; Harry Bruce, "Gardener of the Gulf: The Greening of Alex Campbell and the Preserving of PEI," *Canadian Magazine*, 3 April 1976, 4–8; Lynne Douglas and Martha Pratt, eds., *Energy Days: Proceedings of an Open Seminar of the Legislative Assembly of Prince Edward Island* (Charlottetown, PEI: Institute of Man and Resources, 1976); Donella H. Meadows, Dennis L. Meadows, Jørgen Randers, and William Behrens III, *The Limits to Growth* (New York: Universe, 1972).

28 Science Council of Canada, *Natural Resource Policy Issues in Canada* (Ottawa: Science Council of Canada, 1973), 39; Science Council of Canada, *Canada as a Conserver Society: Resource Uncertainties and the Need for New Technologies* (Ottawa: Science Council of Canada, 1977).

29 Hughes, "The World that Feeds Itself"; Bruce, "Gardener of the Gulf." Amory Lovins popularized the terms "hard technology," "high energy," and "soft technology" during the 1970s, and they were widely used to discuss energy policy during the decade. See Lovins, *Soft Energy*

Paths. Perhaps the best example of the application of Lovins's soft path to Canada is a study carried out by Friends of the Earth for the Department of Energy, Mines, and Resources in 1983. David Brooks, Ralph Torrie, and John Robinson, *2025: Soft Energy Paths for Canada* (Ottawa: Energy, Mines, and Resources, 1983).

30 Hughes, "The World that Feeds Itself."

31 Andy Wells, interview by Alan MacEachern, 27 July 1999, series 11, IMR Fonds, Prince Edward Island Public Archives, Charlottetown (hereafter IMR Fonds).

32 Robert Durie to John Todd, 28 October 1974, Background 1974–1981, series 5, subseries 2, IMR Fonds.

33 Alan MacEachern, *Institute of Man and Resources*, 35–36; Wayne MacKinnon, *Between Two Cultures: The Alex Campbell Years* (Stratford, PEI: Tea Hill, 2005), 244–45.

34 Donald Savoie, *Visiting Grandchildren: Economic Development in the Maritimes* (Toronto: University of Toronto Press, 2006), 85–86.

35 MacEachern, *Institute of Man and Resources*, 62. For a discussion of Canada energy policy, see G. Bruce Doern and Glen Toner, *The Politics of Energy: The Development and Implementation of the NEP* (Toronto: Methuen, 1985); and John Fossum, *Oil, the State, and Federalism: The Rise and Demise of Petro-Canada as a Statist Impulse* (Toronto: University of Toronto Press, 1997).

36 Bruce, "Gardener of the Gulf"; MacKinnon, *Between Two Cultures*, 239.

37 Alexander Campbell, "The Politics of Power," 16 June 1976, PEI Speeches Vertical File, PEI Collection, Robertson Library, University of Prince Edward Island, Charlottetown (hereafter PEI Collection); Alexander Campbell, "An Energy Strategy for Prince Edward Island," 19 January 1977, PEI Speeches Vertical File, PEI Collection.

38 Schumacher, *Small Is Beautiful*, 145–47.

39 MacEachern, *Institute of Man and Resources*, 20–21.

40 See Lynne Douglas and Martha Pratt, eds., *Energy Days: Proceedings of an Open Seminar of the Legislative Assembly of Prince Edward Island* (Charlottetown, PEI: Institute of Man and Resources, 1976). Discussions of "conservation," "energy alternatives," and "energy futures" are of particular note. John Todd, Amory Lovins, and George McRobbie (who worked closely with Schumacher) led the discussion of "energy futures."

41 MacEachern, *Institute of Man and Resources*, 33.

42 Canada-PEI Agreement on Renewable Energy Development, 1 October 1976, Phase II File, 1976–1980, series 3, subseries 4, IMR Fonds.

43 John Todd, "An Ark for Prince Edward Island: A Family Sized

Food, Energy and Housing Complex including Integrated Solar, Windmill, Greenhouse, Fish Culture and Living Components," New Alchemy Institute Vertical File, PEI Collection.

44 United Nations, *UN-Habitat* 12, no. 2 (2006): 4, 16; MacEachern, *Institute of Man and Resources*, 23.

45 Robert Durie, "Technical Review Meeting: Ark for Prince Edward Island," New Alchemy Institute Vertical File, PEI Collection.

46 J. Todd, "An Ark for Prince Edward Island."

47 Editorial, *Eastern Graphic* (Charlottetown, PEI), 25 May 1975.

48 Douglas Smith, "Quicksilver Utopias: The Counterculture as a Social Field in British Columbia" (PhD diss., McGill University, 1978); Miller, *The 60s Communes*, 219–20.

49 N. Todd, "The Opening of the Arks," 14.

50 David Lees, "Aboard the Good Ship Ark," *Harrowsmith*, September/October 1977, 52.

51 Mungall, "Space Age Ark," 103; Susan Soucoup, "Prince Edward Island's Ark," *Harrowsmith*, July/August 1976, 32–35; Lees, "Aboard the Good Ship Ark," 48; Stewart, "The Ark," 42.

52 Lees, "Aboard the Good Ship Ark," 49.

53 J. Todd, "An Ark for Prince Edward Island"; Durie to Todd, IMR Fonds; Durie, "Technical Review Meeting."

54 Mungall, "Space Age Ark," 104.

55 David Catmur to Nancy Willis, "Re: the organization of the Ark" [memo], 4 July 1977, series 5, subseries 2, IMR Fonds.

56 J. Todd, "An Ark for Prince Edward Island."

57 Ron Zweig, "Bioshelters as Organisms," *Journal of the New Alchemists*, no. 4 (1977): 107–13.

58 John Todd, "Tomorrow is our Permanent Address," *Journal of the New Alchemists*, no. 4 (1977): 89.

59 William McLarney, "New Alchemy Agricultural Research Report No. 2: Irrigation of Garden Vegetables with Fertile Fish Pond Water," *Journal of the New Alchemists*, no. 2 (1974): 73–79; William McLarney and John Todd, "Walton Two: A Complete Guide to Backyard Fish Farming," *Journal of the New Alchemists*, no. 2 (1974): 79–117.

60 Angevine, Barnhart, and Todd, "New Alchemy's Ark," 36–37.

61 "The New Alchemists," *Time*, 17 March 1975, 100; Hughes, "The World that Feeds Itself."

62 David Bergmark, interview by Alan MacEachern, 29 July 1999, series 11, IMR Fonds; Durie, "Technical Review Meeting."

63 J. Todd, "Tomorrow is our Permanent Address," 96.

64 J. Todd, "An Ark for Prince Edward Island."

65 Andrew Wells, "Discussion of the New Alchemy Institute's ARK in PEI and the Institute of Man and

Resources," 19 November 1975, IMR Vertical File, PEI Collection.

66 Joe Seale, "New Alchemy Hydrowind Development Program," *Journal of the New Alchemists*, no. 5 (1979): 44–52; J. Todd, "An Ark for Prince Edward Island."

67 J. Todd, "Tomorrow is our Permanent Address," 93–94.

68 Durie, "Technical Review Meeting." Wind energy would eventually come to PEI under the supervision of the IMR. MacEachern, *Institute of Man and Resources*, 43.

69 Wells, interview.

70 Editorial, *Eastern Graphic*, 22 September 1976.

71 Durie, "Technical Review Meeting."

72 Editorial, *Eastern Graphic*, 11 January 1978.

73 MacEachern, *Institute of Man and Resources*, 53.

74 Durie, "Technical Review Meeting."

75 John Todd to Andy Wells, February 1978, series 5, subseries 2, IMR Fonds.

76 Rob Dykstra, "The Ark Sinks," *Atlantic Insight*, August/September 1981, 30–32; Silver Donald Cameron, "Floundering of the Ark," *Maclean's*, 1 June 1981, 13–14.

77 "PEI Ark Demolished," *Guardian* (Charlottetown, PEI), 14 March 2000.

78 Energy Probe, "Ecology House: A Project of the Pollution Probe Foundation," 1 March 1979, vol. 310, file 210, Alistair Gillespie Fonds, Library and Archives Canada (hereafter Gillespie Fonds).

79 See Linda Gilkinson, *West Coast Gardening: Natural Insect, Weed, and Disease Control* (Victoria, BC: Trafford, 2006).

80 Renewable Energy Branch, Dept. of Energy, Mines, and Resources, "Re: Advisory Committee on Conservation and Renewable Energy" [memo], 23 July 1977, vol. 236, file 220-31, 1977, Gillespie Fonds. Todd would never serve on the committee, since the New Alchemists had abandoned the Ark and their Canadian work by the time it began to meet.

81 Alexander Campbell, "Address to the Science Council of Canada," 19 June 1975, PEI Speeches Vertical File, PEI Collection.

82 John Todd and Nancy Jack Todd furthered this transition with the publication of their widely read book *Bioshelters, Ocean Arks, and City Farming* (San Francisco: Sierra Club, 1984).

83 Nancy Jack Todd and John Todd, "Special Report: The Village as Solar Ecology," *Journal of the New Alchemists*, no. 7 (1981): 135–74.

84 James Rakocy, Michael Masser, and Thomas Losordo, "Recirculating Aquaculture Tank Production Systems: Aquaponics—Integrating Fish and Plant

Culture" (Southern Regional Aquaculture Center Publication No. 454, November 2006), http://www.aces.edu/dept/fisheries/aquaculture/documents/309884-SRAC454.pdf; Elizabeth Royte, "Street Farmer," *New York Times*, 5 July 2009. Although the New Alchemists' work is not cited, their integration of aquaculture, greenhouse agriculture, and hydroponics is identical to the "aquaponic" systems discussed.

85 Peder Anker, *From Bauhaus to Ecohouse: A History of Ecological Design* (Baton Rouge: Louisiana State University Press, 2010); Nancy Jack Todd, *A Safe and Sustainable World: The Promise of Ecological Design* (Washington, DC: Island Press, 2005); John Todd, "The Innovator's Sense of Urgency," in Zelov and Cousineau, *Design Outlaws on the Ecological Frontier*, 40–45; J. Todd, "The New Alchemists," 172–83.

Dollars for "Deadbeats": Opportunities for Youth Grants and the Back-to-the-Land Movement on British Columbia's Sunshine Coast

Matt Cavers

Back-to-the-land groups, in the 1960s and 1970s, distanced themselves from conventional authority in a number of ways. Geographically, they moved into rural areas where reminders of the social mainstream—such as disapproving neighbours—would be farther away. They identified and retreated from political evils, as did the large number of Americans who migrated into Canada to avoid the draft. Morally and spiritually, they sought to live "in step with the natural world rather than against it" by setting up rural homesteads and communes where they attempted to practice self-sufficient living.[1] But despite their distrust of authority, back-to-the-land groups were content to take advantage of state funding for their projects when it was available. For a short period in the early 1970s, the Canadian federal government provided modest funding to projects through an experimental youth employment program: Opportunities for Youth (OFY). Some of these funds were granted to a cluster of countercultural back-to-the-land groups based in the Sunshine Coast of British Columbia. The government made these small subsidies as part of a broader strategy to quell social unrest in a turbulent time, and by doing so, they made some unlikely allies in the back-to-the-land movement. On the other hand, the money had unintended effects, such as inflaming the already-tense relationship between counterculturalists and their skeptical rural neighbours.

The Department of the Secretary of State, under Pierre Elliot Trudeau's Liberal government, introduced OFY in the spring of 1971. OFY differed from traditional job-creation programs in that it directly funded projects proposed and initiated by young people. Its unconventional approach reflected the fact that the federal government had created the program not only to reduce unemployment, but also to address growing disenchantment among the youth of Canada. Shaken by the events of 1970—a year that included the October Crisis, riots in Regina and Vancouver, and an unprecedented proliferation of unorganized youth travel—the federal government was anxious to provide "meaningful activities" with which to occupy the nation's youth and thereby cool the social climate. In 1971, OFY represented the largest single expenditure in the federal government's $67.2 million youth employment program, which otherwise included funding for increased hiring in the public service, militia and cadet training, group travel programs such as the Young Voyageurs, facilities for individual travel such as youth hostels, and a handful of other programs and services. In its first year, OFY achieved mixed results.[2]

OFY was widely criticized for having been conceived and put into practice too hastily. Announced in March 1971, by the end of April it had received 8,000 applications for summer grants, of which 2,312 eventually received funding. The number of applications overwhelmed the program's small administration, and in May 1971, its initial allotment of $14.7 million was increased to $24.7 million. In the House of Commons, the parliamentary secretary to the minister of state pointed out that the OFY program had social rather than financial goals, and he boasted that, with the program, "For the first time, a government is financing creativity."[3] But the program faced hurdles

MATT CAVERS

as it began operating: many project grants awarded in the first year arrived late, and many went to projects that the public found objectionable, such as drug counselling services, underground newspapers, and, in British Columbia, communes.[4] In all, just over 8 percent of the total number of applications came from British Columbians; these provided 12.9 percent of the successful submissions and received 9.7 percent of the total funds, some $2.4 million.[5] On the national scale, OFY's legacy is ambiguous. While the program provided valuable start-up funds for many projects that continued after the grants had been spent[6], the federal government's own report on the first year of OFY raised concerns over disorganization in the bureaucracy and the loose criteria that were used to identify successful applications. Subsequent work has noted that its grants—typically up to $1,000 per project employee—were too small to meet the needs of students returning to university in the fall.[7]

This chapter reviews the effects of OFY on a local scale. In British Columbia, over thirty grants were awarded to countercultural back-to-the-land projects, many of which described themselves as communes. Over half of these grants, accounting for nearly $40,000 of government funds, subsidized projects located on the Sunshine Coast, an eighty-some-kilometre stretch of mainland coastline north of Vancouver.[8] This concentration of projects provides a case study, which draws upon personal interviews, OFY project files from Library and Archives Canada, and contemporary newspapers from the Sunshine Coast and beyond. This case contributes to the literature on Canada's counterculture era in two key areas. First, examining a selection of funded projects from a local area allows us to reevaluate some of OFY's successes and failures in detail. For instance, I show that OFY funding decisions were probably made, in this case, on the basis of friendship between applicants and the OFY bureaucracy. Such lapses exposed the federal government to the wrath of the local media. And, while their recipients welcomed them, the grants generated a burst of local hostility that left some of the recipients feeling alienated from their new neighbours. Second, OFY produced a document trail consisting of funding applications and project reports, and its more colourful beneficiaries, such as the Sunshine Coast's back-to-the-landers, drew media attention. In general, the countercultural homesteads and communes of the 1970s left behind scant documentary evidence, but OFY serves as a rare access point into an alternative rural community in Canada, showing us that counterculture ideals popular across North America—such as self-reliant living close to nature and healthy alternative communities—also thrived in this isolated rural area. In contrast to some of the other chapters in this collection, this study illustrates the initial local hostility that greeted the back-to-the-land groups rather than the larger areas of consensus that counterculturalists forged with their neighbours over time. In this sense, the competition for government funds and the public oversight invited by the distribution of state moneys exacerbated existing suspicions.

LOOKING FOR THE BACK-TO-THE-LAND MOVEMENT

British Columbia, by virtue of its reputation for natural beauty and underpopulated land—not to mention its location outside of the United States—drew large numbers of countercultural migrants in the back-to-the-land movement that took shape over the late 1960s and throughout the 1970s. As former Powell River communard Mark Vonnegut put it in his memoir, *The Eden Express*, "Just about everyone, young and old, straights and freaks, wanted to stay up long into the night talking about [looking for land in British Columbia.]"[9] The back-to-the-land movement of the 1970s left its mark on British Columbia's landscape in the geographic distribution of alternative lifestyles, which, to this day, flourish in pockets such as the Slocan Valley, the northern Gulf Islands, and indeed, the Sunshine Coast. Yet for a part of the very recent past—the events discussed here are just more than four decades gone—the back-to-the-land movement in British Columbia is written only faintly on the historical record.

Indeed, it can be challenging to find evidence of the back-to-the-land movement's presence on the Sunshine Coast. For instance, while the area's local history has been written and rewritten a handful of times, "hippies" appear only fleetingly in the two most widely read published accounts.[10] Even the most prolific of contemporary sources, the region's two weekly newspapers, mostly ignored the presence of the newcomers until the OFY grants were awarded in the spring of 1971. This was in spite of the fact the back-to-the-landers would have been clearly visible. The Sunshine Coast in 1971 was home to ten thousand residents, distributed between the villages of Gibsons and Sechelt as well as a handful of smaller settlements such as Roberts Creek, Halfmoon Bay, Pender Harbour, and Egmont. Then, as now, the only way to reach the Sunshine Coast other than flying or boating was to take the car ferry from West Vancouver to Gibsons. In such an (almost literally) insular community, young strangers, many of whom sported long hair, would have attracted attention.

Despite the near-silence of the local press, there are occasional signs from as early as 1967 and 1968 that participants in the emerging counterculture were gravitating to the Sunshine Coast: a classified ad in the alternative Vancouver weekly *The Georgia Straight* announced that a "young married couple" were hoping to find land on the Sunshine Coast "without [the] usual establishment hassle with payments"; the *Coast News* advertised a dance at the Roberts Creek Hall with Vancouver psychedelic rockers Papa Bear's Medicine Band; and the *Coast News* published a letter to the editor by radical Simon Fraser University professors Louis Feldhammer and Margaret Benston, who were outraged at the irony of having themselves been ejected from the Welcome Café in Gibsons for having long hair.[11] Indeed, two veterans of the Sunshine Coast counterculture scene, both of whom arrived in 1969, recalled finding a counterculture community well-established on the Sunshine Coast in that year.[12]

While infrequent, local newspaper coverage of issues related to the countercultural newcomers consistently took a hostile and judgmental tone. Both Sunshine Coast newspaper editors—Fred Cruice of the Gibsons *Coast News* and Douglas Wheeler of the Sechelt *Peninsula Times*—penned editorials as early as 1967 admonishing "hippies" for their alleged criminality and aversion to work. Occasional news stories written prior to the OFY grants reveal the editors' bias against nonconformist young people. For instance, a group that included a local United Church minister approached the town council of Gibsons in 1971 to request support for a summer youth hostel. Council refused to support the proposal, and while both newspapers quoted the town's stridently conservative mayor, Wallace Peterson (he said that "we were never encouraged to bum and if young people choose to travel around the country then they should be prepared to pay their own way"), neither sought the opinions of the hostels' proponents. Cruice mused darkly in his editorial column that "we have problems now, without others showing up."[13]

The local media's otherwise pointed ignorance of the people they called "hippies" came to an end in the late spring of 1971, when the *Coast News* published the names, descriptions, and general locations of nineteen OFY-funded communes and back-to-the-land projects on the Sunshine Coast.[14] This exposure marked the beginning of a turbulent summer for the Sunshine Coast's back-to-the-landers. Before discussing the grants themselves and the controversy they engendered, though, I will address the program that awarded them.

GUERILLA BUREAUCRATS

Jennifer Keck and Wayne Fulks, in a chapter on Trudeau-era youth employment programs, playfully dub OFY's staff "the hip bureaucracy." While it was by no means the first "hip bureaucracy" of the era—the Company of Young Canadians undoubtedly deserves the same recognition, for instance—OFY was administered by a young staff who were well connected in alternative circles. The project officers responsible for OFY grants in British Columbia, employed by the Citizenship Development Branch of the Department of the Secretary of State, enjoyed considerable autonomy from their superiors in Ottawa. In the words of a former OFY bureaucrat, they could "get money into the hands of people who had good ideas and could spend it well," even if their funding decisions caused consternation in the media and on the ground.[15]

Ostensibly, OFY grants were to be given to students returning to school in the fall, "but other young people [were] not excluded" from the competition, and nor were landed immigrants. An OFY brochure explaining "how projects [were] picked," reprinted in the *Coast News*, stated that projects would be selected based on the number of people they employed, the degree to which they involved young people, and whether they created new "services, programs or activities." The brochure advised would-be applicants to obtain application forms from Canada Manpower centres, summer student employment centres, Secretary of State regional offices, Information Canada, and OFY's head office in Ottawa.[16]

On the Sunshine Coast, few of the OFY recipients had to travel far to obtain application forms. In fact, it seems that some program staff actively sought out applicants in the counterculture and cleared the way for their applications. The most effective of these OFY insiders, it appears, was Ken Drushka. Later acclaimed as an author, Drushka worked for OFY in 1971 in a managerial capacity. According to a colleague, Drushka and his friend Colin Thompson were responsible for the majority of the grants that were awarded to back-to-the-land projects in British Columbia. Gregg Macdonald, who worked as a project officer at OFY in the summer of 1971, wryly recalled that by the time he began work early in the summer, "most of the [funding] decisions had already been made on handshakes between Colin Thompson, Ken Drushka, and the wayward youth of the province."[17]

Archival material from OFY appears to confirm this observation. Of the ten evaluation sheets located for this research, seven included Ken Drushka's name under "endorsement or sponsorship." Applicants were clearly aware of the power this name would wield. Robert Morgan, requesting a grant for "Communal Land Development," finished a detailed project description with a non sequitur: "This application was presented to this group by Mr. <u>Ken Drushka</u>." A similar notation, also in Morgan's handwriting, appeared on an application written by his neighbour John S. Gregg, for "Building a Pottery." And when "Richard the Troll," or Richard C. R. Schaller, addressed a personal letter to Ottawa project officer Ian Munro, requesting three separate grants for the activities of the Legal Front Commune, he added in a postscript that "Ken Drushka is familiar with this project." On OFY's internal correspondence, an unidentified project officer noted that the Legal Front was "a group which is central to the youth commune phenomenon on the Sechelt Peninsula" and that they had been "highly recommended by Ken Drushka."[18]

OFY's young staff, in the words of Gregg Macdonald, considered themselves "guerilla bureaucrats," occupying a middle ground between government and counterculture. Macdonald remarked that he had been indifferent to the negative media coverage of OFY, summarizing his position at the time this way: "If our friends in Ottawa want to give money to people . . . whose aspirations we share, and we're confident enough to become the intermediaries, then we're going to do it."[19] In OFY's first year—at least in British Columbia—government employees were unafraid to personally identify with the projects they funded. Thus project officer Tom Ryan, writing to John R. Wimbush of the Legal Front Commune to advise that a late project report had cost the group $50 of their $1,000 grant, explained that "this kind of [project] appeals to me personally so I am rather disappointed that we can't help financially."[20] The OFY staff's role as an intermediary between communes and the government's coffers, rather unsurprisingly, won the federal government the allegiance of some grant recipients. Morgan—a recent immigrant from Seattle—wrote in his project report that "direct Government subsidization of this sort of life-style . . . shows the young people in the country that there is a real chance that some people in the Liberal party, including perhaps the top man himself, actually have some idea about what is really going on out here in the real world."[21]

Not all Liberals, however, looked so kindly upon such "direct Government subsidization." One in particular who was rather more skeptical was Paul St. Pierre, member of Parliament for Coast Chilcotin, the federal constituency encompassing the Sunshine Coast.[22]

MATT CAVERS

PUBLIC OUTRAGE AND PUBLIC SUPPORT

On June 16, 1971, the *Coast News* obtained a list of the OFY grants awarded on the Sunshine Coast and published it alongside an open letter from St. Pierre to secretary of state Gérard Pelletier. In the letter, St. Pierre expressed serious misgivings over the grants, citing "disturbing" criticism from "area residents" and recommending that "municipal authorities and the police" scrutinize the projects. As a Liberal himself, St. Pierre would have been reticent to criticize the government directly, although opposition Progressive Conservative, Social Credit, and New Democratic Party members of Parliament repeatedly queried the minister on the administration of the program and the decision-making process. At the local level, *Peninsula Times* editor Wheeler fumed that "halfwits holding well paid and important positions in a government elected on the Liberal ticket have perpetrated the most outstanding malfeasance ever considered possible." OFY drove Wheeler into such a rage that he spent much of the summer of 1971 hurling editorial missiles at the grant recipients, whom he disparaged as "half the deadbeats in creation."[23]

To what extent Wheeler's vitriol represented the attitudes of his community is uncertain, but on the other hand, it is clear that many people saw his editorials as part of a campaign of intimidation conducted by "the established business community" against "long-haired youth."[24] It was not uncommon, for instance, for "longhairs" to be refused service at local restaurants—an action that Wheeler fully supported.[25] Local politicians displayed animosity toward "hippies," including Gibsons Mayor Wallace Peterson, who said in an interview with a *Vancouver Sun* reporter that "to support a bunch of American hippies . . . isn't the purpose of the OFY program. They're using the money to learn how to grow pot—I don't think you'll see a single potato growing on any of these communes."[26]

Peterson's words spoke to a popular assumption that the OFY recipients, being "hippies," would spend their grant money on the cultivation of marijuana. However, this assumption was never publicly demonstrated on the Sunshine Coast.[27] On June 21, local RCMP officers, along with a party that reportedly included undercover narcotics officers and immigration officials, conducted an unannounced sweep of the OFY projects on the Sunshine Coast. If the police were expecting to find evidence of drug use, as the targets of the sweep believed, they almost certainly failed to do so. Given the editorial stances of the two local newspapers, arrests of OFY recipients would have been reported prominently. The only mention of the sweep, however, was buried in the letters section of the *Coast News* on June 30, in an invited contribution from Ken Dalgleish. A neighbour and friend of OFY recipient Robert Morgan, Dalgleish asked rhetorically, "Was there a suspicion of a crime, or is receiving a grant criminal?"[28]

The OFY recipients were not without local supporters, and the drug raid appears to have encouraged some to speak out. The *Coast News* published a handful of letters defending the program, though some expressed ambivalence; in one letter, Norman Watson hedged that OFY was "cheaper than riot police."[29] In the *Peninsula Times*, columnist Frank O'Brien suggested that OFY was a useful program insofar as it acted as a "safety valve" against youth rebellion, but he scorned grant recipients for overestimating the societal importance of their projects.[30] However, besides a few exceptions such as these, the local press afforded little space to those sympathetic to the OFY recipients.

Negative media coverage and official suspicion of the OFY projects led some of the grant recipients to feel alienated from the community around them, although the depth and duration of this experience varied. Schaller noted in his project report that "the O.F.Y. program was generally put down by various people. . . . The heat cooled off when a form of communication was established." Morgan complained that, while "neighbours and some merchants" supported the members of his project, "antagonism from [the municipal and federal governments] was the one sore point of the whole program."[31] Barbara Yates voiced this feeling most passionately. Awarded a large grant to run a farm retreat for transient girls, Yates wrote in her project report that "there was a lot of open resentment and mistrust towards those who received money, what their motives were and their projects in general, although few people knew any of these 'hippies' personally. . . . We felt, except for several older friends in the community, that we were very alone out here." But Yates, too, acknowledged that as the summer went on, her neighbours began warming to the newcomers in their midst.[32]

By awarding federal grants to a group of people already under public scrutiny, and doing so in an apparently reckless manner, OFY erected barriers between the counterculture and the mainstream on the Sunshine Coast. Several OFY recipients, though, eventually crossed these barriers. Many of them, interviewed in 2011, recalled having mostly positive relationships with all but the most stridently conservative of their neighbours. Of Wheeler, though, all agreed that he was driven by an irrational hatred of "hippies."[33]

The tension between "heads" and "straights" drew media attention from further afield, and some of it differed sharply from the coverage in the local press. The *Vancouver Sun* and *Vancouver Province* each sent reporters to the Sunshine Coast in late June, and by all accounts the reporters were rather more sympathetic to the counterculture than were members of the local press.[34] The resulting feature articles, both published on June 28, identified the OFY grants as a bone of contention. Several of the grant recipients—Schaller of the Legal Front, Morgan of the Crowe Road Commune, and Bill Bradford of the Bayview Commune—were interviewed and quoted extensively, while by contrast, both articles cast local establishment figures in an unflattering light. The *Province*'s Duncan McWhirter noted, "the federal plan and long-haired youth in general [were] being subjected to a fierce attack by Douglas Wheeler." McWhirter wrote that Wheeler had "said . . . that the long-haired youngsters were 'scum, the dregs'" and went on to remark that while Wheeler had been "described by some as a South African, [he] turned out to be an Englishman who had visited the racist, right wing regime [in South Africa] and evidently found much there to please him."[35] Even further removed from the local conflict, the *Montreal Star* and *New York Times* ran feature articles about the Sunshine Coast's communes later in the summer of 1971. Both articles were highly sympathetic toward the projects, with only the *Star* piece making passing reference to any local controversy over the grants. The *New York Times* piece omitted any mention of hostility entirely, beginning on a somewhat incredulous note: "Urban and rural communes have been thriving in this West Coast province for several years. But this is the first summer that some of them have been subsidized by the Canadian Government."[36]

The Sunshine Coast may have been an isolated area with a small population in 1971, but the OFY grants awarded to its participants in the back-to-the-land movement drew an abundance of media attention from local, national, and international newspapers. As I will discuss below, these journalistic sources can give us some sense of what the region's OFY recipients and their neighbours, conservative fears aside, were actually up to.

USING THE CASH

While rural living itself was not countercultural in 1971, what distinguished participants in the back-to-the-land movement from other rural inhabitants was not only their adoption of some form of agriculture, but also their belief in the moral superiority of "the simple life." Rebecca Kneale Gould argues that "modern homesteaders," such as participants in the 1970s back-to-the-land movement, grappled with "a perennial moral problem [which is] that the world as it is today is not the world as it ought to be." By seeking opportunities for honest toil and simple living, then, participants in the back-to-the-land movement engaged in "prefigurative politics."[37] By living simply on the land, they sought to build the foundations of a new society. The summary of the "Basic Organic Greenhouse Gardening" project, based in Sechelt, indicated the group's aims as follows: "To show that organic gardening is a positive alternative to pollution, grow vegetables, to evaluate the effects of chemical fertilization against organic fertilization, and to evaluate the use of pesticides." The group received $950 and claimed that four jobs had been created.[38]

It would be a mistake to see the back-to-the-land movement as politically unified—or uniformly political. Leaving aside the question of whether they constituted a movement, back-to-the-landers varied widely in their motives and their practices.[39] Yet broad themes unified them. As with many of their contemporaries across North America, the Sunshine Coast's back-to-the-landers shared the twin goals of producing food and other necessities and building new forms of community. Indeed, archival and journalistic sources show that while the Sunshine Coast was home to a conspicuous, tight-knit community of people practicing back-to-the-land lifestyles, these people were part of a translocal network of back-to-the-land ideas.[40]

Perhaps the most iconic symbol of the 1970s back-to-the-land project was the rural homestead; indeed, the Sunshine Coast's OFY projects all included at least a minor emphasis on small-scale food production. For some, raising food to become self-sufficient was a primary goal. One group, the Sugar Mountain Commune, received a $1,000 grant to build a "hog shed-barn combination" in which to "produce high grade pork using organic feed." This group's project report is not available, so it is not known how successful they were (or reported themselves to be). However, their hog farm was conspicuous enough to attract the attention of the *New York Times*'s visiting reporter, whose September 1971 story confirmed that there were indeed pigs on the property of grant recipient Henry Rodriguez.[41] On the other hand, the members of the Crowe Road Commune, who received $1,000 to develop their five-acre property, did claim success. Applicant Robert Morgan reported to OFY at the end of the summer that the vegetable garden fed the group well ("largely because we determined . . . how to properly make good compost"), that laying hens and rabbits had supplied occasional protein, and that the group had canned considerable quantities of fruits and vegetables.[42]

Gleaning and recycling were important aspects of self-sufficiency for some of the grant recipients. The Crowe Road Commune group obtained grain for free by sweeping out railcars in Vancouver, gathered unused canning jars by placing classified advertisements, turned fat donated by a local butcher into soap, and somehow "got the right to wreck a house, thereby obtaining a lot of wood of all sorts."[43] This group's creative strategies for obtaining goods might have been more practical than ideological—after all, Morgan notes that the $1,000 grant was "virtually the only support" that the group of twelve adults and four children received over the summer—but others gleaned supplies to avoid harming the land. John Houghton, who obtained a $1,296 OFY grant to build a communal float house, told Rita Reif of the *New York Times* that he would build the house exclusively with wind-felled trees, as "we've destroyed enough of the forest."[44]

OFY also funded small-scale commodity production. Its recipients included, for instance, a candle-maker on a commune in Egmont, a medicinal herb producer on a commune in Roberts Creek, and, as discussed above, a hog farming operation. Unfortunately, reports are available for none of these projects, so journalistic sources provide some of the only details that survive—such as the fact that the residents of the herb farm were unsure whether their business would be viable when their $1,000 grant ran out.[45] However, it is clear that many in this community were interested in developing alternative trade networks. John S. Gregg, a potter, noted in his final report (for "Building a Pottery") that he hoped to "become a potter in some sort of a landscape of people who really need and use pottery in their daily lives because it is made locally and they know the guy that makes it or he helps them to make it for themself [sic]."[46]

One significant venture in this regard was the "general store" created by the Legal Front Commune in Roberts Creek. Intended to market handicrafts, "ecology orientated foods," and local produce, the store was opened on a property that the Coast Family Society, a group representing the local "head community," purchased in June 1971, partially with the proceeds of their OFY grants. Legal Front spokesperson Richard "The Troll" Schaller explained, in a letter to OFY project officer Ian Munro, that the group's intention was to "[set] up a third world economic system, on a small scale." Apparently this involved obtaining produce from local gardens, including some that Legal Front members tended with a separate $1,000 grant. Schaller noted in his project report, though, that "not many of the gardens made it this year and many of the vegetables had to be scored from the longtime commercial organic farms."[47]

MATT CAVERS

Beyond producing food and other necessities, the Sunshine Coast's OFY recipients were involved in community organizing. Local "heads" founded the Coast Family Society, as mentioned above, as an alternative to the mainstream Roberts Creek Community Association, from which they were excluded.[48] To raise funds for the property that housed the general store (and a "people's garage"), the society organized a fundraising picnic on July 17 featuring "electric music and all you can eat for two bucks." Schaller declared the event a success, even though the group only "broke even financially," as only a few of the reported three hundred attendees had paid for admission.[49] While this group existed to serve the counterculture community, another OFY project—the largest on the Sunshine Coast—hoped to build a bridge to the mainstream. The Sunshine Coast Youth Communication and Employment Centre, run by Barbara Yates, operated on a farm near Gibsons primarily as a drop-in centre for transient girls and young women. In addition to providing young people with farm work—which furnished the centre with fresh vegetables—this group offered various services to the broader community, including organizing beach cleanups and providing volunteer labour to the Gibsons Athletic Association. While Yates's attempts at community service were rebuffed at first, as outlined above, the mainstream gradually warmed to the industrious newcomers.[50]

These journalistic and archival sources do not chronicle, but merely suggest, what participants in the back-to-the-land movement were doing on the Sunshine Coast in 1971. Much more took place than was recorded. Many more people passed through the area than stayed, and few of those who had left could be located in 2011 for interviews. Nevertheless, extant sources show that back-to-the-landers on the Sunshine Coast were integrated with broader networks of back-to-the-land ideas and practices. As they did across North America and beyond, the people who went to live on the Sunshine Coast did so to experience "the good life" of community and attempted to achieve a degree of self-sufficiency in growing their own food.

CONCLUSION

While participants in the back-to-the-land movement distanced themselves from authority in many ways, many were quite willing to accept the state's assistance when it was offered to them. Twenty-one federal grants, amounting to nearly $40,000, funded a group of communes and other back-to-the-land projects on British Columbia's Sunshine Coast in the summer of 1971. This chapter has addressed several facets of this complicated relationship between the state, local communities, and the counterculture.

The local newspapers reacted furiously to the grants. They charged that federal funds had been handled inappropriately, which, as it happens, was true. In OFY's first year, individual staff awarded grants to personal acquaintances for projects that politicians and the public questioned. However, the Sunshine Coast's local press went beyond this criticism to attack the grant recipients themselves. *Peninsula Times* editor Wheeler did so flamboyantly, heaping contempt upon the people he labelled "deadbeats" and "hippies." The newcomers had their supporters, but their detractors tended to be people in positions of power, including the local RCMP, the mayor of Gibsons, and the local member of Parliament. As a review of the program concluded, such negative media attention was fairly typical nation-wide: "The initial reaction was almost uniformly critical. Conjecture, and self-fulfilling prophecy produced lurid stories of bungling, depravity, radicalism and drugs."[51] The OFY program highlighted the activities of countercultural groups, creating a space for sustained criticism of their alternative lifestyles. In the face of this public judgment, several grant recipients reported feeling alienated and out of place. On the other hand, while the mainstream may have resented it, these participants in the counterculture heartily approved of the program that supported them, however modestly, in the summer of 1971.

Significantly, the OFY left behind a documentary record of an alternative community in rural Canada. The back-to-the-land movement in Canada has been heavily mythologized (along with the era's other countercultural happenings), but, as elsewhere, the scarcity of documents from back-to-the-land projects makes writing about this phenomenon challenging. OFY, then, briefly shines a light on a period that otherwise might escape our notice. What it reveals to us is that members of a rural community on British Columbia's Pacific coast took part in a search for "the good life" that was carried out all across North America and beyond. Given the opportunity to apply for government financial support, many of their projects reflected keen interests in environmental issues, as seen in their emphasis on food production and organic gardening.[52] And for a brief period, the Canadian federal government provided modest funding to environmental projects that, in the case of British Columbia's Sunshine Coast, reflected the counterculture's partial rejection of mainstream society.

NOTES

1   Rebecca Kneale Gould, *At Home in Nature: Modern Homesteading and Spiritual Practice in America* (Berkeley: University of California Press, 2005), xx.

2   Canadian Council on Social Development, *Youth '71: An Inquiry by the Canadian Council on Social Development into the Transient Youth and Opportunities for Youth Programs in the Summer of 1971* (Ottawa: Canadian Council on Social Development, 1971); Evaluation Task Force, *Report of the Evaluation Task Force to the Secretary of State: Opportunities for Youth* (Ottawa: Department of the Secretary of State, 1972); Jennifer Keck and Wayne Fulks, "Meaningful Work and Community Betterment: The Case of Opportunities for Youth and Local Initiatives Program, 1971–1973," in *Community*

Organizing: Canadian Experience, ed. Brian Wharf and Michael Clague (Toronto: Oxford University Press, 1997), 113–36.

3   Canada, House of Commons, *Debates*, 3 May 1971 (James Hugh Faulkner), 28th Parl., 3rd sess., vol. 5, 5468.

4   Keck and Fulks, "Meaningful Work."

5   Canada, House of Commons, *Debates*, 27 October 1971 (Gérard Pelletier), 28th Parl., 3rd sess., vol. 9, 9077.

6   See, in particular, Ryan O'Connor's chapter in this collection.

7   Evaluation Task Force, *Report*; Keck and Fulks, "Meaningful Work."

8   Secretary of State, *A Canadian Experiment: Catalogue of Projects Funded in 1971 under the Federal Government*

Opportunities for Youth Program of the Department of the Secretary of State (Ottawa: Department of the Secretary of State, 1972); "Grants Continue as Hot Topic," *Coast News*, 23 June 1971. The name "Sunshine Coast" sometimes also includes the Powell River region, but in this paper I consider the Sunshine Coast to be the mainland and islands lying between Howe Sound and Jervis Inlet. This area includes the Sechelt Peninsula, the name of which was sometimes misapplied to the entire Sunshine Coast.

9   Mark Vonnegut, *The Eden Express: A Memoir of Insanity* (New York: Seven Stories, 2002), 28.

10   See Howard White, *The Sunshine Coast: From Gibsons to Powell River*, 2nd ed. (Madeira Park, BC: Harbour, 2011); and Betty Keller and Rosella Leslie, *Bright Seas, Pioneer Spirits: A History of the Sunshine Coast* (Victoria, BC: TouchWood Editions, 2009).

11   *Georgia Straight*, 22 September 1967; *Coast News*, 22 January 1967; Louis Feldhammer and Margaret Benston, letter to the editor, *Coast News*, 22 August 1968.

12   Frank O'Brien, interview with the author, 4 May 2011; Tony Greenfield, interview with the author, 5 May 2011.

13   Minutes of the Regular Meeting of the Council of the Village of Gibsons, 11 May 1971, Gibsons Municipal Hall; "Mayor Cool on Invasion," *Coast News*, 12 May 1971; "Travelling Youth Influx Viewed with Apprehension," *Peninsula Times*, 19 May 1971; "Merchants Warned to Guard Premises," *Coast News*, 21 April 1971; "The Public Youth Program," *Coast News*, 19 May 1971.

14   "Youth Movement Grants Questioned," *Coast News*, 16 June 1971.

15   "James," interview with the author, 27 May 2011. This interviewee recalled that, at twenty-eight years old, he was among the eldest in the Vancouver office.

16   Citizenship Development Branch, "Opportunities for Youth: Notes for Applicants," undated [1971] brochure reprinted in *Coast News*, 16 June 1971.

17   Gregg Macdonald, interview with the author, 27 May 2011.

18   "Opportunities for Youth Program—Communal Land Development—Gibsons, British Columbia," file 500-670 (emphasis in original), vol. 368, RG118-G-3, Library and Archives Canada [hereafter LAC]; "Opportunities for Youth Program—Building a Pottery—Robert's Creek, British Columbia," vol. 360, file 500-814, RG118-G-4, LAC; "Opportunities for Youth Program—The Legal Front Commune—Vancouver, British Columbia," file 500-813A, vol. 368, RG118-G-3, LAC.

19 Macdonald, interview.

20 "OFY Program—Legal Front Commune."

21 "OFY Program—Communal Land Development." Morgan finished his report by inviting government personnel to drop by for a visit, adding, "This includes M[argaret] Trudeau, whom [sic] we understand has relatives in the area."

22 Like Drushka, St. Pierre is better known as an author.

23 Paul St. Pierre, letter to Gérard Pelletier, reprinted in "Youth Movement Grants Questioned," *Coast News*, 16 June 1971; "Media Scavengers Return," *Peninsula Times*, 7 July 1971; "Warm Winds of Change," *Peninsula Times*, 23 June 1971.

24 Duncan McWhirter, "Emotions Running High at Sechelt," *Vancouver Province*, 28 June 1971.

25 "Are We Rural Rubes?" *Peninsula Times*, 7 July 1971.

26 "Sunshine Coast Communes Arouse Local Ire," *Vancouver Sun*, 28 June 1971. According to several interviewees, Peterson was right that many of the OFY recipients were American.

27 However, OFY cancelled a grant to a wilderness camp for teenagers near Prince George after the project's organizers were convicted of marijuana possession on June 11, 1971. "Opportunities for Youth Program—Northern Wilderness Project—Island of Saturna, BC," file 500-571, vol. 378, RG118-G-3, LAC.

28 "OFY—Communal Land Development"; Ken Dalgleish, "Why? Youth Wants to Know," *Coast News*, 30 June 1971. Robert Morgan (interview with the author, 26 May 2011) attributes the police's failure to find illegal drugs to the targets having circulated a warning that the authorities were on their way. "Somebody noticed them on the ferry or something . . . and so the word went really fast."

29 Norman Watson, letter to the editor, *Coast News*, 30 June 1971.

30 Frank O'Brien, "Rushes," *Peninsula Times*, 23 June 1971.

31 "Opportunities for Youth Program—The Legal Front—General Store—Roberts Creek, British Columbia," file 500-813C, vol. 360, RG118-G-4, LAC; "OFY Program—Communal Land Development."

32 "Opportunities for Youth Program—Sunshine Coast Youth Communication and Employment Centre—Gibson's Landing, British Columbia," file 500-43, vol. 355, RG118-G-4, LAC.

33 Bob Morgan, interview with the author, 16 May 2011; Diana Morgan, interview with the author, 4 May 2011; Ken Dalgleish, interview with the author, 2 May 2011. Several interviewees shared colourful recollections of Wheeler. One described him as "more right-wing than Attila the Hun." Another claimed that Wheeler kept a loaded pistol in his desk drawer to defend himself from "hippies," and yet

another that he slept with the pistol under his pillow.

34 "OFY Program—Communal Land Development"; Walter Peterson, letter to the editor, *Vancouver Sun*, 9 July 1971. Peterson began his letter, "I met your 'hippie' reporter you sent up recently."

35 "Sunshine Coast Communes Arouse Local Ire"; McWhirter, "Emotions Running High." Wheeler later retaliated by musing about the "scavengers" and "termites" of the Vancouver media in an editorial ("Media Scavengers Return").

36 Josh Freed, "A New World in Harmony with Nature," *Montreal Star*, 17 July 1971; Rita Reif, "Communes in Canada: Government Lends Hand with Grants," *New York Times*, 15 September 1971.

37 Gould, *At Home in Nature*. The term "prefigurative politics" is borrowed from Wini Breines, "Community and Organization: The New Left and Michels' 'Iron Law,'" *Social Problems* 27, no. 4 (1980): 419–29.

38 Secretary of State, *Canadian Experiment*, 219.

39 Jeffrey Jacob, *New Pioneers: The Back-to-the-Land Movement and the Search for a Sustainable Future* (University Park: Pennsylvania State University Press, 1997).

40 It is difficult to estimate how many back-to-the-landers lived on the Sunshine Coast in this year, but the catalogue of OFY projects provides an extremely rough estimate. Each OFY grant was intended to create a certain number of jobs; adding up these numbers, the total is somewhere between 80 and 150. The uncertainty is due to some of the larger grants being awarded for projects that only partially took place on the Sunshine Coast. Secretary of State, *Canadian Experiment*.

41 "Opportunities for Youth Program—Erection of Hog Shed and Barn—Sechelt, British Columbia," file 500-856, vol. 361, RG118-G-5, LAC; Reif, "Communes in Canada."

42 "OFY Program—Communal Land Development."

43 Ibid.; Diana Morgan, interview with the author, 4 May 2011.

44 "John H.," quoted in Reif, "Communes in Canada."

45 Secretary of State, *Canadian Experiment*; Freed, "A New World in Harmony"; Reif, "Communes in Canada."

46 "OFY Program—Building a Pottery."

47 "OFY Program—Legal Front—General Store." Schaller notes that "eleven of the cash contributors [for the Roberts Creek property on which the store was located] were recipients [sic] of O.F.Y. grants."

48 Duncan McWhirter, "'Heads, Straights' All Getting Uptight," *Vancouver Province*, 29 June 1971.

49 "OFY Program—The Legal Front—General Store," LAC. Schaller also wrote that the picnic "provided an opportunity for the employment of three undercover narcotics officers from Vankouver [sic]."

50 "OFY Program—Sunshine Coast Youth."

51 Canadian Council on Social Development, *Youth '71*, 87.

52 According to the government's own calculations, only 4 percent of all the OFY projects were considered "environmental projects"—but this reflected a narrow definition. Secretary of State, *Canadian Experiment*, n.p. (Introduction).

Building Futures Together: Western and Aboriginal Countercultures and the Environment in the Yukon Territory[1]

David Neufeld

In the twentieth century, the Western world experienced extraordinary growth in its power and wealth. The intertwining of state organization and capitalist economy, while also provoking devastating wars, resulted in stable and prosperous societies promising freedom. By the 1960s, increasing numbers of young people were cashing in on this promise. Both as individuals and as leaders of communities, they challenged conventional notions of social order and sought alternative ways of life. The resulting diversity reflected the notion that freedom entailed not simply doing what one wants in the present, which was certainly popular, but also making a new and different future.

Human relations with the environment are an expression of cultural values and beliefs. In the Western democratic societies of the mid-twentieth century, the romantic counterculture seeking renewed relationships with the environment challenged dominant rational and materialist societal values. The Yukon Territory, with its sparse population and its land widely available for squatting, proved alluring. Incoming back-to-the-landers conceived of the Yukon as untouched wild space, a place where they could build alternative ways of living. They encountered there a different group of people, who had never left the land: the Aboriginal people of the territory, who also sought a future that addressed their interests. The countercultural goals the newcomers brought with them in the 1960s came to fruition in the 1980s. The Yukon counterculture was plural, as both Western and Aboriginal countercultures shaped distinct discourses on environmental relationships. The Western counterculture was interested in getting back to the land, while the Aboriginal counterculture worked to get their land back. Nevertheless, at certain times and on some specific issues, these two countercultural groups cooperated in fashioning alternative futures.

GOING BACK TO THE LAND

One counterculture seeker's Yukon experience illustrates the connection between the back-to-the-land impetus and deep concerns about the environment. Tim Gerberding was born at about the middle of the baby boom.[2] He grew up in Wisconsin, the son of a Lutheran minister. The family's life was disrupted when his father was accused of theological heresy for suggesting that the Bible was a metaphorical guide to living rather than a text of literal truth. Brought before an ecclesiastical court, his father lost his job. His experience of a rigid and righteous organization pursuing a single truth paralleled Tim's later counterculture experiences in the Yukon Territory.

By the time Gerberding was eight, the Lutheran authorities had relaxed their censure, and his father was invited back into the church. A new posting moved the family to Denver, Colorado. A family friend there—the director of the state historical society—took them to many historic sites. The ghost towns and isolated mountaintops of the West appealed to Tim. These trips lengthened, and in 1967 Tim and a friend took off on a summer-long ramble, ending up with hippie friends in San Francisco's "summer of love." Police soon apprehended them as

9.1 Alan Innis-Taylor in his Whitehorse office, 1972. Source: Richard Harrington Coll. PHO 105 389, Yukon Archives.

wayward youth and shepherded them back to Denver, to finish high school.

After two years at St. John's College, a so-called Great Books college, Gerberding and several friends headed off on a road trip. The "buzz" was about "cheap land in Canada," and it was said that the best place was Golden, British Columbia. Inspired by Timothy Leary and other counterculture writers, the friends headed north. Land was available around Golden, but they had nowhere near enough money to buy any of it. Continuing north, they visited the Nass Valley. Tim was overwhelmed by the beauty and isolation of the place; he remembers a strikingly beautiful chunk of land, full of ancient trees, but again the prospective landowners needed more money than they had. Eventually they made it to Whitehorse, in the Yukon Territory. Here they ran into Alan Innis-Taylor, who invited them into his cramped office full of books, maps, and artifacts in the downtown federal government building.[3] He regaled them with stories of the Yukon River and the historic places along it. The group was entranced. But with both the summer and their money waning, they headed south to regroup and make plans for a return to the Yukon.

The following spring, Gerberding and four young male and female companions bought an old school bus, loaded it with supplies, and headed back to the land. In Whitehorse they again met with Innis-Taylor, now noticeably cooler about encouraging young people to go off into the bush. He emphasized that the Dawson area was an especially poor choice. Gerberding and the others figured that Innis-Taylor's warnings must mean Dawson was an especially interesting place. With a full set of topographical maps of the middle Yukon River, they marked all the promising places: that is, those with southern exposure, likelihood of dry timber, a side stream for clean drinking water, and a good place to build a cabin. At Pelly Crossing, they built a log raft and launched their search. The five young people floated downstream. The seemingly endless rolling hills and empty forests of the Yukon valley flowed past them during the truly never-ending sunny days of a subarctic summer. Steering into likely spots, they wandered through their selections of "free" land, dreaming about what they could do. Their experience was an almost mythic idyll of the counterculture.

Toward the end of summer, they floated past the confluence of the Forty Mile River, some ninety kilometres downriver from Dawson. Coal Creek, across the river, was the last place marked on their maps. They landed. A recent forest fire had left a lot of standing dead trees suitable for a cabin and firewood. On the flat land beside the river a rough airstrip had been cut to support a fire camp. A large tank of diesel fuel still sat beside the strip, and on the high bank above the river there was a beat-up wooden trailer to live in. After a few days' stay, they decided it would be home. The group retrieved their bus and supplies, driving to the Clinton Creek mine townsite, just ten kilometres upriver from "their land." With the help of another young couple camped out on the river nearby, they ferried their gear over the river to their new place. Their romantic adventure in the backwoods of the Yukon was, for some of them, almost over; however, for Gerberding it was just beginning. Determined to stay and live in this apparently pristine environment, he felt in control of his future.

POSTWAR YUKON AND ECONOMIC DEVELOPMENT

The Yukon had long inspired such reveries of escape from mainstream society. From early in the twentieth century, the North inspired Canadian thought. The North's resources and its demands upon the human spirit were seen as the promise of a bright national future. However, only the Yukon, of Klondike gold rush lore, and the railway belt depicted in Group of Seven paintings had any purchase on Canadian popular culture. Through mid-century, it was war—hot and, later, cold—that sparked a more concrete Canadian attention to the defence capacity and natural resource wealth of the North. The Yukon Territory consequently experienced considerable change as the modern state and industry began directing and constructing the envisioned national future. These changes eventually both spawned and supported the diverse set of counterculture responses in the Yukon beginning in the mid-1960s.

The wartime Alaska Highway and CANOL (Canadian Oil) Road, both military projects completed in 1943, and the postwar expansion of the road network connected the Yukon to the outside world.[4] This enhanced transport access along with a variety of government incentives supported more intensive mineral prospecting. Mining activity in the Yukon accelerated through the 1950s. Production of copper restarted at the Whitehorse mines after World War II; the short-lived Johobo copper mine began operations within the recently established Kluane Game Sanctuary in 1959; and the large asbestos mine at Clinton Creek, not far from Gerberding's homestead, was under development by 1964. In 1969 the huge Cyprus Anvil lead/zinc open pit mine started operation, resulting in the new town of Faro.

Even grander visions of the future built on the almost unimaginably large hydroelectric power generation opportunities in the Yukon. As early as 1946, the Aluminum Company of America proposed a hydro project in the upper Yukon basin to support aluminum production.[5] Another proposal suggested that the entire upper watershed of the Yukon River be reversed, to flow south to the Pacific Ocean. In

1949 the US Bureau of Reclamation suggested that the scale of such a hydroelectric project might require the town of Whitehorse to be moved, arguing that while "local residents . . . would resist such a move . . . [this] should not influence the planning of the project for the national good of both Canada and the United States."[6] As elsewhere in the resource periphery, outside desires trumped local perspectives.

Successive Canadian governments agreed, celebrating the national prosperity generated by the mining and hydroelectric industries. Under the direction of Progressive Conservative Prime Minister John Diefenbaker's "Northern Vision" of national development and progress, small-minded local opposition should not hinder progress. In 1960, Gordon Robertson, deputy minister of northern affairs and natural resources, summed up Canada's position: "We own the north. . . . It belongs to us. Canadians for this reason, must look to the north to see what it is good for, to see how to use it."[7] Annual northern development conferences, bringing together federal geologists and bureaucrats, industrial venture capitalists, and the northern business community, started in the mid-1960s.

The application of this attitude from the 1950s through the 1970s was especially virulent in the Canadian North because of the assumed absence of any local countervailing philosophies of social order. Traditional Aboriginal societies were pushed off balance by the colonial administration exercised by the federal government. The postwar newcomers created a transient, unstable community prepared to accept overarching and dehumanizing social ordering in the belief that the Yukon was too big to hurt and in return for generous personal material gain. An observer in the early 1970s noted, "The notion that the territory's wilderness environment is infinite and that it somehow constitutes either a loss or a threat to society is evident both in individuals' interactions with their surroundings and in the aggressive governmental programs designed to open up or conquer the frontier."[8]

Although they may be considered back-to-the-landers in their own right, Rudy Burian and Yvonne Burian exemplify those non-Aboriginal old-timers unconcerned about big industrial plans. The Burian

Camp is: A place that does not really exist. It has no history and no future. It has no plans and no memories...
Camp is: A place where you keep your mouth shut because you learn that you have nothing to talk about except camp...
Camp is: A meat grinder for your soul: it swallows you, grinds you up and delivers you to someone's plate as their workhorse...
Camp is: A beginning of a lifestyle you hope to move into – i.e., school, money for another start on the land...
Camp is: Where you give up your freedom of choice for a solid helping of chance...

9.2 A view of the transient experience in the Yukon. Source: Rock & Roll Moose Meat Collective, *The Lost Whole Moose Catalogue: A Yukon Way of Knowing* (Whitehorse: Rock & Roll Moose Meat, 1979), 98. The *Lost Whole Moose Catalogue*, through its three distinct editions (1979, 1991, and 1997), provides a fascinating record of an arriving, ageing, and next-generation Yukon counterculture.

family lived on Stewart Island, an isolated outpost on the Yukon River, at about the midway point of Gerberding's raft trip. Until the early 1950s the island had a roadhouse and a small store, a police post and a telegraph station. River shipping ended with the construction of the gravel road to Dawson, and the Burian family soon had the island to themselves. The Burian land holdings on the island were not freehold, but a lease of ten of the fifty-by-one-hundred-foot lots of the Stewart River townsite plotted in the fall of 1899 that had never been developed. Despite the lack of land security, the Burians remained sanguine about the threat of large-scale mining development: "I never worry about that, 'cause that's about all it's good for up here is mining.

That's what keeps the country going. That's what the Yukon is."
Commenting on the landscape devastation resulting from sixty years
of gold dredging about Dawson, Rudy noted, "Yeah, but in a few years
you'll never even see that. It will be just the way it was. . . . There's just
too much land for it all to disappear like it does outside. It might happen sometime. But not in our lifetime."[9] Like many non-Aboriginal
Yukoners, the Burians accepted resource development and could not
imagine it significantly changing the Yukon environment.

Ultimately, a revival of First Nations political activity in the 1960s
and the arrival of counterculture youth in the 1970s challenged the
prevailing pro-development approach. Aboriginal people quickly saw
how such changes would impinge on their lives. While many Yukon
non-Indigenous people welcomed, or at least accepted, economic
development, government and industry actions significantly compromised Yukon Aboriginal peoples' relationships with the natural
world. In 1947, the territorial council—made up exclusively of non-Indigenous men—revised hunting regulations to address the interests of
local sport hunters and to broaden access to wildlife for both tourism
development and big game outfitters. This desire to maximize the economic value of wildlife resulted in much stricter controls on access to
the land. Until this time, Aboriginal access to wildlife had been largely unregulated, the government accepting subsistence practices as a
positive alternative to relief payments. The new regulations, however,
applied to all, both Aboriginal and newcomer. With the expansion
of the federal social safety net in the postwar period, even isolated
groups were guaranteed their subsistence needs.[10] Waged jobs were
available for the progressive, welfare for the reluctant. Them Kjar, the
first director of the Yukon's Game and Publicity Department, wrote
with satisfaction about these changes in 1954:

> If we look back only five or six years we find the times in the
> Yukon have changed greatly due to the many new mining,
> prospecting, and building enterprises which suddenly have
> been established, as well as improved road and air trans-
> portation, thereby enabling trappers (Indian and White) to

occupy themselves elsewhere at a much higher profit than trapping or hunting could give, leaving obsolete the old way of living off the country as well as nullifying the use of dogs.[11]

Others were less enthusiastic about such decisions. Jack Hope, a New York writer investigating Yukon peoples' responses to these changes in the early 1970s, noted the challenges faced by Aboriginal people: "Another destabilizing aspect of Yukon society is the collision between the territory's white and Indian cultures. . . . These problems are further exacerbated by the white culture's typical intolerance for a people who could not make a smooth and instant transition from a relatively primitive society to a modern industrial one."[12] Tr'ondëk Hwëch'in elder Percy Henry responded to my interest in First Nations perspectives on the counterculture newcomers by reminding me that the hippies were not the only young people in the region worth noting.[13] He spoke of his own youth. When a young man needed money he just headed off into the bush. Henry recalled he would set up camp in a good spot and cut wood for a week or two; then, hauling it into town, he'd sell it and have money. But then things changed: "every piece of land has a number on it." He could no longer just go out in the bush. What were young people supposed to do? "Regulation, regulation, regulation, halfway to Heaven."[14] This pressure on Aboriginal land and resources occurred at the same time and in the same place that Gerberding and his friends were laying out their camp in the woods, escaping contemporary society—or so they thought.[15]

The Aboriginal challenge to contemporary society took a different form. Establishing a number of activist organizations in the mid-1960s, Yukon Aboriginal people organized themselves to confront the government's vision of the future. They wished to define their own relationships with their land, within the cultural landscape they called home. While there should be no confusing First Nations' intercultural resistance with the intracultural protests of the Western counterculture, there were places and times where their distinct strategies and different objectives intersected.[16] Both, however, related to the character of peoples' cultural and social relations with the environment.

The Yukon in the 1970s was a difficult place from which to challenge the contemporary world. A small long-term non-Aboriginal population ran the commercial and administrative infrastructure of the territory. The bulk of newcomers were simply sampling life in the North, making some money for a project back down south or starting their career in government. New York writer Hope was struck by the casual alienation of most of the white people he met:

> The highly transient nature of the Yukon's population is not conducive to social stability. The territory's frontier economy is based on construction and resource exploitative occupations, such as mining and mineral exploration, road and dam building. These occupations offer extremely high wages to attract men to the remote, outpost locations, but they do not encourage roots.[17]

Hope also noted "people who appear each spring . . . to see what the frontier is all about. . . . [M]ost go back south at about the time the weather turns cold and the days get short. Some are back the next June with a zealous Yukon patriotism and a fierce determination to stick out the next winter. A few do. Most don't."[18] All Yukoners—newcomers, old-timers, and Aboriginal people alike—faced steep odds in countering the power of a centralized government's push for economic growth.

WORKING ON THE ENVIRONMENT: THE YUKON CONSERVATION SOCIETY

However, a small number of enthusiasts started a spirited, quixotic intracultural resistance to the excesses of the government's northern vision. The story of the Yukon Conservation Society (YCS) highlights the nature of both the countercultural desire to limit environmental devastation and the conflict with the local non-Aboriginal population,

DAVID NEUFELD

who saw only the promise of modernity and doubted that there could be any serious threat to the territory's expansive wilderness.

John Lammers, a refugee of the World War II Nazi occupation of the Netherlands, arrived in the Yukon in the early 1950s. Originally settling in Whitehorse, Lammers undertook a variety of bush and town positions until 1963, when he acquired land at the isolated confluence of the Stewart and Pelly rivers and set up a year-round wilderness tourism business. He and his wife built their own camp and ran river trips for a small but well-to-do market of southern Canadians and Americans. Their income was modest, but Lammers was living out his dream of an alternative lifestyle. He lamented the fact that many newcomers simply settled in town and adopted a suburban lifestyle, when the alternatives were so attractive:

> The physical Yukon is different from elsewhere. And with planning, our society up here could easily offer human beings a life that is different. But to do that we would have to . . . acknowledge that the thing that is special about the Yukon is her small population, our space, our great natural environment. And our society should steer people toward a lifestyle that takes advantage of her particular endowments. . . . There are many, uniquely Yukon opportunities.[19]

Years before the influx of counterculture youth, Lammers identified the Yukon as a place that could address the Western cultural interest in communing with the natural world.

Lammers's lifestyle aspirations quickly ran into the realities of the Yukon mining boom of the mid-1960s. Incensed, Lammers complained that local citizens had no

> voice in the planning of what goes on. . . . The federal Department of Indian Affairs and Northern Development . . . rules the Yukon . . . [controlling] oil exploration, road building, timber [and] mining. And they apparently view their function as one of a . . . broker, selling off our product

A Castle on the Frontier

9.3 Lammers' autobiographical book showing the Yukon Wilderness Unlimited camp at mouth of the Pelly River. "It's a wonderful place here. It's friendly. I have always felt the wilderness hospitable and warm. It's more than just the physical facts of water and trees . . . it's sort of a medium, like amniotic fluid that surrounds the child in the womb and invokes a feeling of total well-being. We feel good here. And it means something to me to have built my home here, as carefully as I could, to fit into the wilderness." Source: Hope, *Yukon*, 157.

DAVID NEUFELD

as fast as they can, without trying to ration any of it out to last for the future.[20]

Lammers started a citizens' campaign for comprehensive land-use regulation. As president of the new YCS, established in 1968, he wrote, "We are in danger of losing all of the Yukon's natural assets swiftly, if greedy, single-minded, unplanned, extraction type of 'development' is allowed to spread its cancer here also."[21] In late summer 1970, speculators staked Lammers's own property for potential development.[22] Lammers moved into high gear.

The society, closely modelled on the Alaska Conservation Society,[23] was led by local outdoorsmen and -women. These included Charlie Taylor, the president of the Yukon Fish and Game Association; Monty Alfred, a federal hydrologist; Bob Charlie, a young First Nations broadcaster; and Cora Grant, an avid birdwatcher and the one stalwart supporter of Lammers's causes. Lammers began to build the society's membership, gaining the support of the local canoe club and the consumers' association; the chamber of commerce and all government departments studiously ignored them.

An initial survey of the membership identified subjects of concern: wildlife preservation, scenic and aesthetic aspects of Yukon roads, cleanup of abandoned mines, public consultation by the federal government, public education on issues, and parks and land-use regulation.[24] These relatively conservative objectives reflected Lammers's desire to support the federal government's proposed introduction of comprehensive land-use regulations that the local mining industry vociferously resisted. Lammers had difficulty getting the YCS board to support even these limited goals. Membership was never large; he complained that only ninety people signed up, and over sixty of these were from southern Canada and the United States.[25] Among local members, only two or three stood with him on more controversial issues. One by one, directors resigned or simply stopped showing up. Rudy Burian, Lammers's downstream neighbour, observed that "[John] wants to save everything. He even believes in suing the government if they do something he doesn't like. He's a nice guy, but he's just too radical in his conservation ideas. . . . His ideas to me are more or less communistic."[26]

The failure of the first conservation society to advance an environmental agenda among Yukon people can be attributed to the prevailing non-Aboriginal belief in the scale and resilience of Yukon wilderness. In 1971 roughly three-quarters of the Yukon population—largely young, non-Aboriginal, and transient—lived in Whitehorse or the relatively large communities of Dawson City, Faro, and Watson Lake. Caught up in the glamour of a new Klondike rush, they did not see how they were connected to contemporary environmental issues.[27]

However, Kluane National Park, an integral element of the government's northern development strategy, garnered all kinds of interest. Southern environmental organizations, the National and Provincial Parks Association of Canada being especially prominent, rallied broad public support for the establishment of the Yukon national park.[28] Although originally supportive, Lammers found that his agenda for land-use regulation was lost between the economics of industry, the symbolic value of the national park, and the economic diversification offered by tourism.[29] Isolated and almost alone, he concluded the federal government had traded away the regulations to industry in return for the national park. A bitter man, Lammers attempted to disband the YCS in the spring of 1972.

Despite Lammers's fiat, the conservation society carried on. The floundering group was briefly led by a non-Aboriginal believer in ecological salvation through Native spirituality. However, the outdoorsmen and more conservative long-time Yukoners quickly took over leadership and pursued a more moderate public role. They made contacts in forward-thinking elements of the mining industry, and together they sought to fashion compromises in mining practices. They were no dreamers of an alternative future. As its new president declared, "Conservation . . . must make the leap from dreamy Indian *idyll* to present day *push*. . . . Members of the society can create a working relationship between the simple life and today's life."[30]

At this point, the counterculture reacted in their own way to environmental threats. Under the leadership of Innis-Taylor—"the grand old man of Yukon environmentalism"[31]—they established an alternative body, the Yukon Resource Council, in 1973 to maintain a strong public voice against unrestricted resource development. The council soon recaptured the leadership of the YCS. They mounted potent professional and technical cases against proposed mega-hydroelectric projects, the Alaska Highway pipeline, extension of the Dempster Highway, and the related release of lands for oil exploration. A teacher in Old Crow (a YCS member) supported the Vuntut Gwitchin community presentation to the Berger Commission (1974–1977), an early crossover between counterculture and Aboriginal advocacy. Further, during the anti-trapping and anti-fur campaigns of the mid-1980s, YCS was almost alone among Canadian environmental groups in offering support for Aboriginal trapping.

By the late 1970s federal government departments, now more sensitive to demands for local participation, began to support the YCS with annual grants and specific consultation contracts, much to the chagrin of the local Progressive Conservative MP, Erik Nielsen.[32] The society's environmental education role greatly expanded in the early 1980s. Programs were developed for schools, and a much broader offering to the public included workshops on energy conservation, lectures, and a series of travel books highlighting Yukon's environmental wonders. These efforts, especially the initiation of a still-operating summer program of free nature and history hikes in Whitehorse, dovetailed with the development of the ecotourism market. The organization was well organized, employed paid staff, and enjoyed a degree of community support. Led by a board of well-educated and articulate wilderness guides, teachers, and professionals—most of them young recent arrivals in the territory—their strategic objective was the transformation of Yukon society.

In 1979, YCS President Nancy MacPherson noted that the society wished "to explore and promote alternative ways of thinking and living in this world." Lynda Ehrlich, an active member in 1980, recalled local resentment toward the society: "YCS was perceived as kind of radical left wingers and [the YCS] wouldn't have disputed that to a great extent. . . . There were all sorts of crazy comments about us, the hippies."[33] The society was radical. Most of its activities promoted rethinking humans' relationship with nature and argued for a reduction in resource consumption and a greater emphasis on the stewardship of natural places.

While the work of the society was non-partisan, its membership was not. Politics in the Yukon was then, and largely remains today, polarized between the business and industry promoters of unrestrained economic development and a counterculture recognizing a plurality of interests in how the environment is understood and related to. In 1985 the two Yukon countercultures felt they had achieved a major objective with the election of a left-leaning New Democratic Party government with four First Nations and four non-Indigenous legislators under the leadership of Tony Penikett. This victory was understood as a sign of the transformation wrought by both First Nations young people and their newcomer peers over the previous fifteen years.[34] Many more of them subsequently moved into government to enact their dreams.

AN ABORIGINAL COUNTERCULTURE

The period between World War II and the mid-1980s witnessed a dramatic assault upon Yukon society. Prior to the war, First Nations were generally left to their own devices and ways of life. While economic development occurred, its scope was generally limited in areal and environmental effects. Local government generally tolerated the different ways of life practiced by First Nations. The intrusion of big government into the Yukon during and following the war radically transformed this situation, and for at least a half century, the values of a callous modernity seeking material wealth and a homogeneous national society were forced upon an unwilling Aboriginal population. The Yukon is still recovering from this onslaught and its lingering agents.

The Tr'ondëk Hwëch'in of the Dawson City area have identified the effects of successive government actions as causing three separations: from their land, between generations, and from their history.[35]

The separation from the land began in the early 1940s with more aggressive federal land management. The creation of the Kluane Game Sanctuary in 1943 as a national park reserve challenged the viability of a number of surrounding First Nations communities.[36] The concurrent revision of land-use and hunting regulations resulted in a loss of Aboriginal young people's personal agency.[37] For a people whose way of life, both material and spiritual, relied upon an intimate relationship with land, this separation was a major crisis.

Linked with the creation of the national social safety net in the late 1940s was an expansion of the Yukon Indian residential school system. Community church schools, perennially underfunded, were closed, and more children were removed from their families and subjected to an education that undermined their certainties, replacing them with foreign values. The resulting separation between generations shattered the community's ability to flourish, excising a sense of purpose and isolating parents and elders from their future. The residential schools absorbed the young people of the counterculture generation and spawned in many of them the same restless energy that activated their non-Aboriginal peers.

Beginning in the mid-1950s the tourist and public promotion of a Canadian history of the Yukon erased Aboriginal people from time. A focus on the incorporation of the Yukon into Canada and the exploitation of natural resources for the nation altered the earlier non-Aboriginal narrative of gold discovery as the catalyst for a self-governing progressive community.[38] While the original community story was a narrative that ran parallel to an Aboriginal presence, the national revision in place by the late 1960s was a totalizing narrative that denied any other stories of presence. This corruption effectively removed First Nations from the Yukon landscape and compromised their ability to make their interests known. These three traumatic separations seriously tested the resilience of their communities. Unlike their Western counterculture contemporaries, Aboriginal young people sought to overcome the restrictions on their use of their own lands and then to make their own future in a culturally plural Canada.

Yukon First Nations, still very much present, responded to these pressures by initiating both a fight *for* freedom (i.e., direct negotiation with government, court action, and public protest) and a fight *of* freedom (i.e., working within their communities without reference to the limitations of colonial laws).[39] They centred their fight *for* freedom upon obtaining a treaty with Canada, an alteration of the national thinking by an appeal to Western traditions of law and social justice. Through the use of state tools, First Nations hoped to achieve their objective of national recognition and respect for their cultural presence in Canada. Yukon First Nations leadership, raised through the social turmoil and distress of the 1940s, 1950s, and 1960s, began the fight for freedom by preparing a proposal for government consideration. In early 1973, they presented to Prime Minister Pierre Elliot Trudeau a document titled "Together Today for Our Children Tomorrow: A Statement of Grievances and an Approach to Settlement by the Yukon Indian People."[40] This document challenged the denial of the place of Yukon First Nations in Canada and proposed a settlement. The First Nations' objectives were to regain their connections to their land, restore their cultural relationship to the environment, and establish self-government. Together these would provide the capacity to build an alternative future for their people. The subsequent negotiation and implementation of the treaty, elements of which are still in progress, have taken almost fifty years. The process, often bitter and confrontational, has absorbed the lives of three generations of First Nations people. The ultimate outcome of this fight for freedom is still in the balance.[41]

Alongside the fight *for* freedom was the fight *of* freedom waged within communities. Communities struggled to renew traditional values and land practices as part of the rejuvenation of their cultural identity. Effective self-government requires a people who know how they are related to their environment. Typical among Yukon Indian bands in the 1960s, the Dawson band council (now the Tr'ondëk Hwëch'in government) strove to protect their community, shielding it from the colonial excesses of Department of Indian Affairs (DIA) programs. A thankless and crippling responsibility, leadership in this

Together Today
for our
Children
Tomorrow

by the
YUKON INDIAN PEOPLE

9.4 Cover of *Together Today for our Children Tomorrow.* "We had our own God and our own Religion which taught us how to live together in peace. This Religion also taught us how to live as part of the land. We learned how to practice what is now called multiple land use, conservation, and resource management. We have much to teach the Whiteman about these things when he is ready to listen. Many Indians look at what the Whiteman has done to destroy and pollute lakes and rivers and wonder what will happen to the birds, fish and game. We wonder how anyone will be able to know what effect the Pipeline and other industrial projects will have on birds, fish and game before they are built. We feel that you are going ahead to build the Pipeline anyway, regardless of the harm it will do. . . . We wonder how the Whiteman can be so concerned about the future by putting money in the bank, and still he pays no attention to the future of the land if he can make a quick dollar from selling it to foreigners. Traditionally the Indian did not have to store up goods for the future, because he protected the resources so that they would always be there." Source: Council for Yukon Indians, *Together Today for Our Children Tomorrow*, 9, 14–15.

period took its toll on the participants. Nevertheless, this resistance allowed community activities and structures to continue operating despite the many outside forces seeking to modernize the Tr'ondëk Hwëch'in.[42] The gradual move of families from Moosehide, the Tr'ondëk Hwëch'in village just downriver from Dawson, to Dawson through the 1950s undermined community coherence and strength. In the later 1960s, the band council encouraged people to visit their former home. Moosehide quickly became a sanctuary from the pressures of assimilation. People regained their spirit through maintenance of the graveyard, an opportunity for youth and elders to work together, and the repair of their homes and the village church, pillaged by non-Indigenous river travellers. Moosehide was an anchor to place and signalled community agency. The fight of freedom also acknowledged the need to live together with non-Aboriginal peoples in the present. Early efforts to work with some counterculture young people showed the way. Tim Gerberding's experiences illustrate the character of contact between Aboriginal and newcomer.

Having branched off from the friends with whom he had come to the Yukon, Gerberding's first years were busy as he and his partner developed a satisfying subsistence lifestyle. They built a permanent cabin, established a large vegetable garden, harvested berries, and began fishing the annual runs of salmon. They were not alone; a small number of other young newcomers were scattered along this remote stretch of the Yukon River. They shared the work during salmon runs and supported one another in times of need. They also had contact with local First Nations people. Initially, there was tension. (What are these young strangers doing in my trapline and hunting area? And what are they doing fishing for salmon?) For the Tr'ondëk Hwëch'in, the presence of another growing group of earnest and aggressive newcomers was unwelcome news.[43] Competition for salmon, whether for subsistence or commercial sale, fuelled conflict. However, the situation mellowed as prolific salmon runs through the 1970s and 1980s supported the needs of all.[44] First Nations families also relaxed as only a small number of young people stayed for long in the area—and the newcomers were generally polite. They also appeared to prefer camps far upriver, a comfortable distance from the Tr'ondëk Hwëch'in fam-ily camps closer to Dawson.[45] With patience shown by First Nations and respectful approaches made by the newcomers, accommodation became possible.

By 1980, both newcomers and Tr'ondëk Hwëch'in fishers came to acknowledge some shared interests. The establishment of a viable commercial salmon fishery that year demonstrated the possibilities of a sustainable life on the land, something both the First Nations and the back-to-the-landers wanted. When Dawson and Old Crow First Nations, with DIA support, built the Hän Fisheries plant in Dawson in 1982, the back-to-the-landers upriver were active, and welcome, contributors to the success of the operation.[46] Gerberding and his family moved into Dawson in the late 1980s as their two boys ap-proached school age, satisfied they had lived the bush life. Gerberding started work with the Tr'ondëk Hwëch'in government as a member of its treaty negotiating team—a professional relationship that continues to the present.

A growing sense of a brighter future encouraged Tr'ondëk Hwëch'in young people to further action. In the mid-1980s, a number of mothers took action to rebuild the community's traditional connec-tions to the land. Concerned about their children losing their identity and with families effectively being confined to town, the women bad-gered their brothers—since children's uncles are traditionally respon-sible for teaching land skills—to get their children out on the land. The result was First Hunt. The initial event quickly became a com-munity affair, and First Hunt continues today. Focused on teens, and open to both Aboriginal and non-Aboriginal youth, the hunt is tied to the arrival of the caribou herd in Tombstone Territorial Park, north of Dawson City. The school closes, and youth, First Nations elders, and the "uncles," now including a cross-section of Dawson Aboriginal and non-Indigenous hunters, set up a large camp. Activities include hunter safety, environmental science games, elder storytelling ses-sions, hunting, and, usually late in the evening with Coleman lanterns hissing yellow light inside a large canvas-wall tent, the butchering of caribou carcasses. A couple of weeks later the young hunters host a community feast where they share their first kills with the whole community. The success of this adaptive reproduction of the traditional Tr'ondëk Hwëch'in annual round was subsequently expanded to include First Fish at Moosehide in July, beaver camp in the spring, moosehide tanning in later fall, and a variety of gathering activities that vary with the seasons. Each provides close contact among youth, elders, and extended families, exercising and reinforcing the important connections between generations, the environment, and their lands.

From the mid-1960s to the present, Tr'ondëk Hwëch'in young people, with concerns typical of Yukon First Nations, pursued an intercultural countercultural agenda emphasizing their distinctive cultural attachment to the environment—a connection expressed in stories and song, in land-use practices, and in their history. In a profound sense, the countercultural agenda of the First Nations peoples of the Yukon rejected the tenets of economic growth proposed by governments, corporations, and many Yukon individuals. The First Nations preferred to open a space for the unimpeded development of their own cultural, social, and economic interests and values. They did so, in part, in collaboration with the people who had left the mainstream of North America in search of a new way to build a future.

CONCLUSION

In the 1960s and 1970s, young and generally well-educated newcomers came to the Yukon seeking an alternative way of life. Self-sufficient and eager, many squatted in the bush, built their own cabins, planted gardens, and attempted a subsistence lifeway, often with mixed results. Others lived in town and took jobs but held similar values in terms of the environment—specifically, that the environment deserved acknowledgement and care. Together, and with some old-time Yukoners, they began to question the frantic pace of resource extraction and their apparent inability to be heard. The YCS provided a platform for the articulation of protest, to challenge national ideas of the North as Canada's future, with the dramatic environmental changes that industrial activity implied. Once firmly established, the YCS worked most effectively within a familiar Western discourse of nature as a source of both human wealth and solace. Advocating for protected areas and more effective land-use regulation, the Yukon newcomer counterculture remained within a familiar political and cultural realm, seeking an appreciation of nature and a respect for the ecological mechanisms that ensure a healthy environment. Their enthusiasm contributed to the election of the NDP territorial government in 1985.

Yukon Aboriginal peoples, increasingly separated from their lands by government through the mid-twentieth century, faced a very different fight to maintain their connection to the environment. First Nations addressed the intercultural conflict of interests with a dual strategy of diplomatic negotiations with Canada (a fight *for* freedom) and a community-based strategy of adaptation (a fight *of* freedom). With a culturally distinct relationship to the environment, Yukon First Nations struggled to frame their connection to traditional territories in ways that could be understood by Canada's negotiators who wished to understand the land as property. As there is, as yet, no common understanding of the parameters of the Yukon First Nations' relationships with the environment, the long-term validity of the present treaty—the set of agreements and implementation schedules signed between 1992 and the present—remains uncertain. As with the intracultural arrangement wrested by the Western counterculture, the Yukon First Nations intercultural agreements are similarly volatile and remain open for continuing negotiation.

The Yukon countercultures, both newcomer and Aboriginal, reacted to the excesses of the modern colonial administration and the aggressive capitalist economy through the second half of the twentieth century. They pursued different objectives and developed their own tactics. However, both held to the idea that whatever their relationship to the environment might be, the two groups would have to live together and share that environment between them. The diversity inherent in cultural pluralism, and its possibilities of multiple futures, demands a respect and appreciation of fellow travellers.

NOTES

1 I am indebted to many Yukoners for sharing both their research and their personal experiences. The work of historical researchers Will Jones and Gail Lotenberg established a solid foundation for my interpretation of the countercultures in the Yukon. Both Gerry Couture and Tim Gerberding generously shared their memories of settling on the Yukon River in the early 1970s, while elders Ione Christenson and Phyllis Simpson, who grew up on the Yukon River in the 1930s and 1940s, and Linda Johnson, retired Yukon archivist, kept me from making up stuff about the old days.

Tr'ondëk Hwëch'in elders Percy Henry, Mabel Henry, Peggy Kormendy, J. J. Van Bibber, Angie Joseph Rear, Ronald Johnson, Julia Morberg, and John Semple, among many others, have patiently guided my work with their community, while Tr'ondëk Hwëch'in citizens Debbie Nagano, Gerald Isaac, Freda Roberts, Edith Fraser, Georgette McLeod, and Kylie Van Every regularly questioned my ideas and suggested alternative ways of thinking things through. Sue Parsons, Jody Beaumont, and Glenda Bolt, cultural heritage employees of the Tr'ondëk Hwëch'in government, always welcomed me into their offices.

I am also grateful to Colin Coates, the participants of the Counterculture and Environment workshop on Hornby Island, and two anonymous peer reviewers for their suggestions for improving my original paper.

2 Tim Gerberding, interview with the author, 22 February 2011.

3 Innis-Taylor (1900–1983) was born in England, was raised in Canada, trained with the Royal Flying Corps in World War I, served as a Mountie in western Canada, and ended up in Whitehorse in 1926. He subsequently took to mining and later worked as a purser on the Yukon River boats. His Yukon experience made him a suitable chief of operations for Byrd's Antarctic expeditions in the 1930s. During World War II he was an officer in the United States Air Force, supervising weather observations and air-rescue services. In the postwar period he provided Arctic survival training for Western air forces and airlines and developed Arctic survival equipment. In his later years he returned to the Yukon, undertaking conservation education, recording historic sites, working toward the establishment of the territorial archives, and providing advice to all comers from his office in the federal building. He was particularly interested in the welfare of young people coming into the Yukon. http://arctic.synergiesprairies.ca/arctic/index.php/arctic/article/view/2173/2150.

4 The highway to Mayo and Dawson, and its Top of the World

extension to the Alaska boundary, was completed in 1955.
In January 1959, intermittent construction of the Dempster Highway began to support oil field development; construction was eventually finished in 1979. The Robert Campbell Highway, originally completed as a mining road to Tungsten, NWT, in the early 1960s, completed a connection to the Dawson Highway for the Faro mine in the late 1960s. In the mid-1980s a road extension south to the Alaskan port of Skagway replaced rail haulage for the Faro mine. Since then, no significant additions have been made to the Yukon road system.

5   Claus M. Naske, "The Taiya Project," *BC Studies*, nos. 91–92 (Autumn/Winter 1991): 5–50.

6   Quoted in ibid., 20.

7   G. Robertson, "Administration for Development in Northern Canada: The Growth and Evolution of Government," *Journal of the Institute of Public Administration in Canada* 3, no. 4 (1960): 362.

8   Jack Hope, *Yukon* (Englewood Cliffs, NJ: Prentice Hall, 1976), 8.

9   Ibid., 50–51. Conflation of Rudy and Yvonne Burian's statements.

10 Catherine McClellan with Lucie Birckel, Robert Bringhurst, James A. Fall, Carol McCarthy, and Janice R. Sheppard, *Part of the Land, Part of the Water: A History of Yukon Indians*

(Vancouver: Douglas & McIntyre, 1987), 94.

11 Quoted in Gail Lotenberg, "Recognizing Diversity: An Historical Context for Co-managing Wildlife in the Kluane Region, 1890-Present" (manuscript report, Parks Canada, 1998).

12 Hope, *Yukon*, 9.

13 Percy Henry, interview with the author, 25 February 2011.

14 Percy Henry, personal communication, Spring 1996.

15 Gerberding spoke about the ease of getting Yukon land around Dawson in the early 1970s. All one needed to do was identify the place, register their interest in the Lands Office, and, eventually, obtain a survey of the plot—at a significant cost. He paid $1,200 to get the survey done a number of years after settling.

16 Sherry Smith, *Hippies, Indians, and the Fight for Red Power* (Oxford: Oxford University Press, 2012), 7.

17 Hope, *Yukon*, 8.

18 Ibid., 9.

19 John Lammers, *A Castle on the Frontier* (Salt Spring Island, BC: Gray Jay, 2004); Hope, *Yukon*, 155.

20 Lammers, quoted in Hope, *Yukon*, 156.

21 "The issues . . . and a course of action," YCS *Newsletter* (Whitehorse), no. 1 (Fall 1968), 3.

22 Lammers, *A Castle on the Frontier*, 429.

23 The Alaska Conservation Society was established in 1960 at the University of Alaska Fairbanks to lead a citizens' charge against both Project Chariot, a nuclear bomb engineering project promoted by Edward Teller, and the Ramparts Dam on the Yukon River. Dan O'Neill, *The Firecracker Boys: H-bombs, Inupiat Eskimos, and the Roots of the Environmental Movement* (New York: St. Martin's, 1994).

24 Will Jones, *A History of the Yukon Conservation Society: Focus on Kluane National Park 1968–1992* (Whitehorse: Yukon Conservation Society, 1997), 11–12.

25 Many of the non-Yukon members had been clients of Lammers's wilderness tours. Hope, *Yukon*, 154.

26 Quoted in ibid., 51.

27 The national historical commemoration of the Klondike gold rush beginning in the late 1950s was prompted by a government desire to highlight a northern vision of economic development. David Neufeld, "Parks Canada and the Commemoration of the North: History and Heritage," in *Northern Visions: New Perspectives on the North in Canadian History*, ed. Kerry Abel and Ken Coates (Peterborough, ON: Broadview, 2001), 45–75.

28 The National and Provincial Parks Association of Canada, a citizen group dedicated to raising the profile of protected areas, was established in 1963 with significant financial support and overall direction from Parks Canada. It was renamed the Canadian Parks and Wilderness Society in 1986.

29 Jones, *History of the Yukon Conservation Society*, 13.

30 "Editorial," YCS *Newsletter* (Whitehorse), no. 4 (May 1973), 2; emphasis in original.

31 Jones, *History of the Yukon Conservation Society*, 31.

32 Resources—Lands, National Parks, Natural Resources Council, file 3, vol. 180, series 1, Erik Nielsen Fonds, Yukon Archives, Whitehorse.

33 Lynda Ehrlich, interview by Will Jones, Research Collection for Yukon Conservation Society, Parks Canada, Whitehorse. Even as recently as the early twenty-first century the remains of this old guard of hippie-hating anti-environmentalists lingered in the Whitehorse office of the Federal Mining Recorder. I had arranged with the Mining Recorder to exchange a Yukon mining history book I'd published for their digitized inventory of air photos. When I arrived in the office and introduced myself to the counter staff person, she not so amiably replied, "Oh, you're that asshole from Parks."

34 "The NDP Rise to Power in the Yukon," *The National*, CBC Television, broadcast 14 May 1985, http://www.cbc.ca/player/Digital+Archives/Politics/Provincial+and+Territori-

al+Politics/Yukon+Elections/
ID/1803621162/.

35   The three separations and their effects were identified by a Tr'ondëk Hwëch'in community committee in a proposal to the Aboriginal Healing Foundation in the early 1990s. The author worked with the group as secretary.

36   For details of the effects of the sanctuary's creation on the Southern Tutchone community of Burwash Landing, and the community's response to separation from their lands, see David Neufeld, "Kluane National Park Reserve, 1923–1974: Modernity and Pluralism," in *A Century of Parks Canada, 1911–2011*, ed. Elizabeth Claire Campbell (Calgary: University of Calgary Press, 2011), 235–72.

37   More details on these regulation changes are available in Lotenberg, "Recognizing Diversity."

38   David Neufeld, "Public Memory and Public Holidays: Discovery Day and the Establishment of a Klondike Society," *Going Public: Public History Review*, no. 8 (2000): 74–86; David Neufeld and Linda Johnson, "Local History of the Yukon," in *Encyclopedia of Local History*, 2nd ed., ed. Carol Kammen and Amy H. Wilson (Lanham, MD: American Association of State and Local History, 2012), 587–89.

39   James Tully, "The Struggles of Indigenous Peoples for and of Freedom," in *Political Theory and the Rights of Indigenous Peoples*, ed. Duncan Ivison,

Paul Patton, and Will Sanders (Cambridge: Cambridge University Press, 2000), 36–59. For more details on the Tr'ondëk Hwëch'in fight of freedom, see David Neufeld and Georgette McLeod, "Call and Response: The Tr'ondëk Hwëch'in Dänojà Zho Culture Centre: A Canadian First Nation Statement of Cultural Presence" (paper presented at the Northern Governance Policy Research Conference, Yellowknife, November 2009), abstract at NGPRC website, accessed 5 January 2013, http://ngprc.circumpolarhealth.org/abstracts.

40   Council for Yukon Indians, *Together Today for Our Children Tomorrow* (Brampton, ON: Charters, 1977), accessed 5 January 2013, http://www.eco.gov.yk.ca/pdf/together_today_for_our_children_tomorrow.pdf.

41   Numerous negotiations on treaty implementation continue, and several court cases launched by Yukon First Nations seek to clarify the opportunities and possibilities under the Umbrella Final Agreement between Canada, the Council for Yukon Indians, and the Yukon Territory. The Tr'ondëk Hwëch'in alone has pursued at least two cases: one during their negotiations in the mid-1990s and a current case testing the terms of the land-use planning chapter of the agreement.

42   This section is based upon numerous personal communications from Percy Henry and Peggy Kormendy, both of whom

were young chiefs of the Dawson Indian Band Council in the 1960s and 70s.

43 On a joint First Nations and Fishers' Association project, see Dan O'Neill, *A Land Gone Lonesome: An Inland Voyage along the Yukon River* (New York: Basic Books, 2006), 18–29. On the earlier days of contact between back-to-the-landers (a group he calls "back-to-the-bushers") and Native Americans in the Eagle area of Alaska, see John McPhee, *Coming into the Country* (New York: Farrar, Straus & Giroux, 1976), part 3.

44 Jody Cox, "The Upper Yukon River, The Salmon and the People: A History of the Salmon Fisheries" (draft manuscript report, Parks Canada, Whitehorse, April 2000), 94.

45 Ibid., 94n273. Cox points out that the allocation of campsites and fish eddies on the middle Yukon River is an informal affair (pp. 97–98). A map from the late 1970s in the land negotiation files of the Tr'ondëk Hwëch'in archives shows the camps and fishing areas identified by the First Nation for their citizens, mostly close to Dawson, thus allowing others to use areas not identified as such, further downstream from town.

46 Ibid., 94.

DAVID NEUFELD

10

Nature, Spirit, Home: Back-to-the-Land Childbirth in BC's Kootenay Region[1]

Megan J. Davies

Born in 1952 in Honolulu, Pamela Stevenson came to the University of Victoria as an undergraduate student in 1974, but she did not remain long. The following year she took her tuition fees and bought a horse, a kiln, and five hundred pounds of clay; with her husband she then made her way to the Slocan Valley, where she found "Wilderness, mountains and rivers without end." She believed that within "an incredible community of compassionate, educated urban refugees . . . [there] was no better place to raise a family." Four years later, Pamela's daughter, Tara Mani Stevenson, was born at home in Winlaw with the help of Abra Palumbo and Pat Armstrong, two unregistered community midwives in the region. Family photos of the birth celebrate Tara's first day of life, as she was taken to visit the garden that her parents had created, wrapped in a handmade community baby quilt sewn by her grandmothers and local friends.[2]

10.1 Nature and homebirth: Pamela Stevenson takes a walk in her Winlaw, BC, garden with her newborn daughter, Tara, 1978. Source: Stevenson photo collection.

Women like Pamela Stevenson were part of a radical redefinition of childbirth in Canada and the United States during the 1970s and 1980s. Potential parents allied with sympathetic health practitioners to create a sustained critique of the standard hospital birthing procedures of the 1950s and 1960s, which they regarded as having pathologized and medicalized a natural process of the female body.[3] The social movement engaged in the "new midwifery" and homebirth projects reconstructed birth not as a medical event, but as an important life moment that spoke to the natural, collective, female, and spiritual aspects of reproduction and that should take place not on the maternity ward, but in a familiar home setting.[4] As was the case at other counterculture locales, such as The Farm in Tennessee, homebirth in the Kootenays represented an embrace of organic life

MEGAN J. DAVIES

processes and the world of nature.[5] In each of these locations, men played a supportive role, but women spearheaded the grassroots push for change in birthing practices.[6]

Interconnected themes of nature, community, and home thread through Stevenson's story of the birth of her daughter and other Kootenay homebirthing tales from the 1970s and 1980s. Videotaped interviews with these mothers and their midwives are my primary source material for this research, but I also use personal photographs, textile art, documents that homebirth advocates and midwives produced, and books that midwives and their clients consulted.[7] The oral testimonies that I collected are both narrative constructions of the lives of a group of counterculture women and a documentation of their daily experiences. They capture important subjective experiences not otherwise accessible.

The story of homebirth in the Kootenays illuminates the role of counterculture women in creating cultural and social capital and constructing alternative identities.[8] Revisiting the emergence of counterculture homebirth through the lens of environmental history helps bring key themes of place, nature, and maternalism into focus. Historian Carolyn Merchant would recognize Stevenson's triumphant trek through the garden with her newborn daughter, delivered at home without recourse to medical technology, as a recovery narrative, demonstrating how counterculture women reclaimed nature through the act of birth.[9] Homebirthing back-to-the-land women may or may not have read Herbert Marcuse's *One Dimensional Man* (1964), but their praxis demonstrates that they were questioning the fundamental notion of scientific progress, rejecting institutional bureaucracies and health technologies, and reconceptualizing the application of medical knowledge. Their understanding of safe and appropriate natal care moved them to create an underground system that was community- and female-centred, non-invasive, preventative, and pluralistic. Homebirth and midwifery—and women in the counterculture more generally—consistently bisect the human/nature dichotomy that, according to Gregg Mitman, environmental historians need to interrogate.[10]

The homebirthing women of the Kootenays whom I interviewed were interested in "quality of life" environmental issues; for them, the environment encompassed their bodies, their homes, their children, and their food. Like the bearded men on the ship *Phyllis Cormack* bound for the island of Amchitka on Greenpeace's first protest, homebirth advocates were also deeply political, taking counterculture women to the same margins of legality as early environmental activists. In Canada, midwifery was alegal (outside the law) until the 1990s, and homebirthing women were keenly aware that if they or their baby died, their midwife or their family could be charged with manslaughter.[11] The decision to have a homebirth may therefore have appeared to be a personal choice, but the risks it entailed—and the mediation of those risks—were collective and political rather than individual and private.

This chapter begins by situating Kootenay homebirth and midwifery within the broader counterculture history of the 1970s. I then consider the back-to-the-land homestead and the birthing body as political and cultural sites for reclaiming childbirth. The third section discusses how homebirth fostered the formation of alternative counterculture identities through multiple avenues of social and cultural expression. The "natural" and the "homemade"—that which derives from the home space or the home community—emerge as key motifs for homebirth in the Kootenays, motifs designed and executed by counterculture women. Underpinning this narrative is an analysis of how marginal places and practices hold particular value as potential sites of resistance because of their physical and ideological distance from the mainstream. This interpretation therefore posits place, praxis, and the spirit as being of equal importance in fostering radical social and personal change.

IN THE CONTEXT OF SOCIAL MOVEMENTS

The late 1960s and the 1970s were a period when alternative culture and politics flourished in Canada's westernmost province. This movement resulted in a wide-ranging set of initiatives, which included small socialist and feminist presses, radical newspapers, co-op radio stations, radical health groups, communal housing and food co-operatives, experimental education, art, and architectural design projects. While no definitive profile exists of the men and women involved in these schemes, or in the broader movement that gave birth to them, most were young, educated, middle-class, and white. Numerous participants migrated to British Columbia from the United States and other parts of Canada, including thousands of young American draft resisters and others who protested their country's involvement in the Vietnam War by crossing its northern border.[12]

This search for alternative lifestyles had a strong rural orientation.[13] Like Mark Vonnegut, who moved to an isolated commune north of Powell River, British Columbia, in the early 1970s, many who questioned dominant political and social mores relocated to the countryside because of concerns regarding urban pollution, alienating work processes, and the high cost of city life. Vonnegut captured the radical, experimental mood of the back-to-the-land movement in his book *The Eden Express*: "We expected to get closer to nature, to each other and our feelings ... and develop entirely new ways of being and experiencing the world ... free of the cities, of capitalism, of racism, industrialism, they had to be for the better."[14]

Young back-to-the-landers settled on the islands and in rural communities around the Strait of Juan de Fuca and in scattered settlements throughout the province. Yet in many ways the Kootenay region was the quintessential BC rural retreat, holding near-mythical status within the counterculture with its green mountains, clean water, and cheap land. People came to the area from California, the Prairies, and central and eastern Canada to establish farms and live closer to nature, rejuvenating communities where the population had been in decline since the late 1950s.

The emergence of homebirth and midwifery in places like Argenta, Nelson, Kaslo, and the Slocan Valley was part of a broader project of creating an alternative community.[15] Like their urban sisters in Vancouver, Kootenay homebirthing women were also involved in creating non-traditional schools, new food systems, and environmental organizations like the ones chronicled by Nancy Janovicek and Kathleen Rodgers.[16] The feminist critique of birth as a site of stolen female power made sense to this group; many of them had personally experienced male physicians as paternalistic, judgmental, and condescending.[17] Adopting the countercultural use of the body as a site of rebellion against mainstream society and rejecting the idea that nudity was reprehensible, they accepted the body as "natural."[18] Most of my interviewees were also attracted to non-invasive, alternative health therapeutics and aware of childbirth reform work being done in other locales by renegade physicians, radical midwives, and parents searching for natural birthing options.[19] When Pat Armstrong posted a notice in Winlaw in 1971 advertising childbirth education classes, she found a ready clientele.[20]

These new regional residents, many of whom had been raised in suburban or urban settings, saw the Kootenays as an unspoiled and healthy space where they could recapture the lifestyle and values of an imagined rural past. But unlike urbanites who may envision a bucolic existence in an otherwise sanitized countryside, the back-to-the-landers anticipated that their engagement with the rural environment would be productive, even elemental.[21] They understood self-sufficiency and home production as key aspects of the movement. Earthy, "traditional," healthy, and clearly home-crafted, birth in a Kootenay A-frame house fit neatly into the 1970s back-to-the-land project.[22]

HOMESPACES AND THE NATURAL

Home is central to the childbirth stories I collected and was fundamental to re-scripting female reproduction as a counterculture event. Memories of the experience were inevitably peppered with evocative descriptions of the last trip to the outhouse, the dishes left on the table from the night before, the unfortunate choice of the loft bed in which to birth. The decision to deliver in a dwelling rather than in a hospital served to frame counterculture birthing within the long-standing cult of domesticity and multiple ideologies that cluster around and impinge on the act of birth—beliefs about motherhood, the family,

MEGAN J. DAVIES

fatherhood, and parenthood and childhood.[23] Yet the counterculture home—associated with good health, "place identity," security, and privacy—was also a staging site for bringing nature, the female body, maternalist culture, and the community back into the birth process.[24] The back-to-the-land "homeplace" therefore functioned simultaneously as a location where the act of birth was honoured as traditionally female and familial and as a revolutionary space where birth, nature, and motherhood could be radicalized and rendered political through resistance to mainstream medical dominance.[25]

Referencing a historical time when women gave birth at home, and home was central to the circle of life, homebirth also introduced the element of choice in birth, a key aspect of alternative health movements of the period.[26] Pamela Stevenson told me, "Home is a temple. And when you have your child at home with the music you want, and the candles you want, and the pace you want, and the people you trust more than anything . . . it is completely unviolated, sacred space."[27] A natural, drug-free homebirth reclaimed traditional use of the home as a space where important life transitions took place and reworked the postwar family home in countercultural terms. The doctor, the institution, and, by extension, the state were excluded from an event that welcomed father, family, midwives, and friends.[28]

This process had important gendered implications. The close identification of home with self that was evident among back-to-the-land women, and the work they did in producing homeplaces, echoes the lives of their mothers; "home" was an extremely powerful theme in the years following the disruption of World War II, when many Canadian families struggled to find places to live.[29] As Elaine Tyler May makes clear in her work on American families during the Cold War era, the external and internal organization and appearance of family dwellings were important characteristics of postwar society.[30] The Kootenay women transposed this intense identification with home onto a very different kind of living space, a place where "the natural" and the home-crafted were central.

The women who shared their stories with me believe that being born in the rural back-to-the-land home, so profoundly part of the land on which it rested, grounded a baby in a deep physical sense, creating an identity that was intimately connected to place. Abra Palumbo named her second child Forest after the woods on Hornby Island where she was born. "My daughter still lives on this tiny piece of earth that she was born on," Slocan Valley resident Lisa Farr told me, "and that is part of her connectedness to this land and this place and her family."[31]

Back-to-the-land women engaged in many types of labour that were not traditionally female—most notably tree-planting—yet their work as producers of children and re-creators of the natural clearly placed them in conventional female positions in counterculture households. Journalist Myrna Kostash found that counterculture women were expected to do the bulk of the daily work in the home, cooking, caring for children, and managing tasks such a laundry, often without electricity, running water, or the labour-saving devices their mothers had enjoyed.[32] More recently, scholars have emphasized the essential gender conservatism of the movement: the domestic role of women in the counterculture was essentially subservient. Historian Gretchen Lemke-Santangelo has argued that the counterculture normalized heterosexuality and was "deeply committed to so-called feminine traits and values."[33] "Natural" food preparation, as detailed in West Coast counterculture cookbooks such as *Laurel's Kitchen* and *Earth Market Cook Book: Recipes for a Simple Life,* involved labour- and time-intensive processes such as canning, grinding wheat, kneading bread dough, and making yogurt.[34] American counterculture historian Warren Belasco rightly notes that this kind of slow-food preparation was in itself a meditation and a protest against "modernity," but he fails to acknowledge the labour of counterculture women in creating the natural and the nostalgic.[35] In fact, there are clear parallels between making natural food and giving birth naturally: both were processes that claimed a link to older, pre-industrial modes of living. But women's roles and responsibilities made this connection even clearer.

Yet the history of counterculture homebirth can also be interpreted as a place where women claimed power, albeit in traditional spheres.[36] Known in the region as "women who were not afraid of blood or organic stuff, not fearful of the birth process," the Kootenay community midwives were figures of knowledge and authority and were skilled at creating spaces undefined by biomedicine and institutional control—safe home situations where women could know and own the profound moment of birth.[37] Trained through a local apprenticeship system by experienced colleagues and radical physician Carolyn DeMarco, midwives such as Barbara Ray and Abra Palumbo brought a hybridity of alternative and biomedical techniques to the counterculture homebirth.[38] Like the other midwives, Camille Bush had blue cohosh and angelica root on hand for assistance in expelling the placenta, ginger compresses for perinatal support, and shepherd's purse to help control bleeding.[39] But the midwives also knew the mechanics of birth and were competent in the use of standard medical procedures including sterile technique, blood and urine testing, and pelvic measurement.[40]

Just as counterculture nudists rejected the puritanism of churches, schools, and media in freely displaying their bodies, homebirthing women resisted the authority claimed by medicine to manage their bodies. Routine hospital delivery regimes of the period included shaving pubic hair, insisting that women labour only on their back, relegating the prospective father to a waiting room, using hospital gowns and drapes to mask the body, and employing an anesthetic that obliterated the woman's memory of the moment of birth. Some hospitals even strapped women in labour to the delivery table.[41] In contrast, the counterculture home offered women the opportunity to reject entirely these medical models and to labour and birth where and how they chose, surrounded by intimate friends and family. The midwives were women whom a pregnant woman would know from the local daycare or health food store, and a closer relationship between the two women would be forged over the months of gestation.

DeMarco's photograph album chronicling births she attended reveals these elements.[42] The gaze on the naked female body in DeMarco's album is neither clinical nor sexual, but celebratory, for

10.2 Back-to-the-land birthing environs: Woman in labour in the loft of a Kootenay cabin with midwife and female friend in attendance, 1973. Photograph by Jean Hanley Wells. Source: Private photo collection.

the images commemorate the female body in the labour of birthing or focus on the moments after the baby has safely arrived.

While birth is clearly the central topic in this collection of images, home, the world of women, and the natural setting are key framing devices. In one series, a woman in heavy labour stands on a porch in the sunshine. Behind her, a woman is applying pressure to the pregnant woman's lower back, likely to alleviate the pain of back labour. To her left and very close is DeMarco, and to her right, with an arm cradling the labouring woman, is yet another woman, long hair swept back in a braid.

Like homebirthing manuals of the era, these photographs demonstrate that women's bodies were honoured during a homebirth, and the sense of shame that imbued the naked form in the post–World War II era expunged.[43] While environmental history literature on nudism is limited, I link the presentation of the body in DeMarco's album to

MEGAN J. DAVIES

10.3 Homebirth as a female-centred event: Sylvie Lafrance in labour in her home by the Slocan River, 1984. Source: DeMarco photo collection.

Marguerite Shaffer's analysis of the egalitarian and natural unclothed body at American nudist resorts of the postwar period, though my research on the homebirthing body suggests a need to rework her understanding of counterculture nudism as personalized political performance.[44] In one image from DeMarco's collection, a heavily pregnant woman stands smiling, naked, in her garden. Another sits unclothed on the crossbeam of a house under construction, nursing a tiny newborn infant. In a photo that appears to have been taken shortly after birth, a nude man and woman lie in bed, a tiny baby tucked close to the woman's breast.

The counterculture home was the pivotal site for the homebirth-ing work of women in the movement. Nature, evoked through a celebration of the maternal body and the history of home as a site for female-centred birth, was a key component of the argument for homebirth. This linkage between nature, the home, and the birth, however, must be appreciated as radical and traditional at the same time.

HOMEBIRTH AND THE FORMATION OF IDENTITY

For the counterculture women whom I interviewed for this project, the creation of alternative identities through homebirth was both a personal and a collective process, shaped through an affinity with organic processes and an appreciation for the special capacities of women. Relocating childbirth from the hospital to the homestead, these women emphasized their self-reliant, healthy practices and capable bodies; the personal growth that accompanied homebirth; and their engagement in collective cultural practices that acknowledged and celebrated birth as a social and spiritual event. Clearly regarding homebirth as rooted in the local social and physical environment of the region, the Kootenay Parents, a group formed in 1981 to advocate for childbirth choices, argued that homebirth deserved respect as part of parents' culture in the Kootenays.[45]

Birthing in the counterculture Kootenay homestead was integral to a whole set of individual and collective alternative life choices, connecting a longer tradition of maternalist thought and identity with liberation philosophies of the era.[46] Leslie Campos stated, "For me midwifery was about personal choice and personal power. Everything you do, you do naturally—whole food, kerosene lamps, outhouses. . . . When you look to the medical or educational system or any system to have all the answers for you, I think you end up with less freedom."[47] Campos, who moved to the Slocan Valley in 1976 and gave birth to several children at home in the following years, points toward the way in which personal and collective empowerment, self-reliance, and nature converged in counterculture homebirth. Ellie Kremler's oral history makes similar connections. Kremler came to the Kootenays in pursuit of a pioneering lifestyle, seeking "an intense participation in the life process," by growing her own food and running the local co-operative. Homebirth, she told me, was part of this larger picture.[48]

Like many other women and men who migrated to the Kootenays in the 1970s, Lisa Farr described the move as a transformative life experience, interpreting her decision to have a natural homebirth in 1981 as an aspect of the construction of an alternative individual identity:

> To be somewhere where I could be myself . . . there is something about personal growth and becoming a bigger version of yourself—this is my choice, this is me having this baby and taking responsibility for the choices and who I become through [the process] that is really important.[49]

Accordingly, counterculture homebirth was about taking greater personal responsibility for health, claiming control of one's body, individual identity, and hence destiny—hallmarks of the late-twentieth-century alternative health movement and the broader back-to-the-land ethos.[50] Alternative identities were formed through self-reliance, spirituality, and the bringing of nature (or the natural) into daily life. Many of the Kootenay narratives that I collected bear a striking similarity to the typology of illness in Arthur Frank's quest narratives, wherein the "sick" person gains the ability to be self-reflective and proactive, often undertaking heroic battles with the unenlightened in the process.[51] But counterculture homebirth differs both in the integration of nature and politics in the process and in the collective nature of the enterprise. The larger counterculture community provided important practical and emotional scaffolding for this course of action, and community midwives aided prospective parents in preparing for the unique birth they were "seeking and creating."[52]

The quest elements of the Kootenay homebirthing narratives demonstrate that place and a back-to-the-land lifestyle afforded birthing mothers an identity as physically strong and capable women. Local midwives echo this perspective, typically describing their early clients as self-reliant, well educated, and extremely physically fit. Community midwife Pat Armstrong reported that her back-to-the-land clientele from the 1970s were in excellent shape: "They ate out of their organic gardens, chopped firewood, and helped build their own homes. A Kootenay woman would not necessarily shy away from shingling a roof at seven months pregnant."[53] Several midwives that I interviewed linked the lack of complications in pregnancy and childbirth among Kootenay counterculture women to the fact that these women were "highly motivated, highly educated, awake and aware... living on the land, eating well."[54]

Although not all claimed the feminist label, Kootenay women who chose homebirth understood their actions to be part of a larger movement in which women gained control over their own bodies. While only a few of the women that I interviewed identified their actions specifically as feminist, most employed the language and ethics of the movement, emphasizing the importance of choice and the element of power inherent in reclaiming birth from the medical establishment.[55] Here again, we can see how maternalist understandings fuse with a radical contemporary ideology—in this case, feminism. After her first delivery at home, midwife Armstrong was "infused with power" and "knew that women had to take control."[56] Susan Vetrano, Liz Tanner, and Diane Holt, mothers who lived and gave birth in Argenta, all agreed on the importance of women having the power to read their own bodies and make their own choices in pregnancy and childbirth.[57]

For homebirthing women in Argenta, Nelson, Kaslo, and the Slocan Valley, crafting identities as responsible, self-reliant, and personally powerful people meant self-education: learning as much as possible in an experiential fashion about the bodily processes of pregnancy and childbirth. Counterculture childbirth education worked on individual and collective levels, rendering expert knowledge accessible to a lay clientele in a fashion typical of the broader women's health movement and of alternative social movements of the period more generally.[58] Pregnant women, their partners, and their children attended biweekly prenatal clinics run by the midwives, first at the house of Dr. DeMarco, then at the Vallican community hall, and later at the old South Slocan schoolhouse, midway between Nelson and Castlegar.[59] Expectant mothers would come for a morning or afternoon visit; all would attend a shared lunch with a visiting speaker and a discussion group. Posited as natural, the pregnant female body was thus knowable to an audience unschooled in medicine. Midwife Camille Bush noted as well the importance of the clinics as purposeful efforts at building community, an illustration of how homebirthing worked to foster social capital in the counterculture community of the Kootenays.[60]

Health historian Mike Saks identifies the countercultural critique of scientific medicine as essentially anti-modern, discarding notions of rational progress, objective medicine, and professional expertise. At first glance, this characterization should fit the Kootenay homebirthing mothers perfectly, but it does not. At the biweekly clinics, pregnant women connected with a hybrid form of midwifery practice that encompassed both the biomedical and the alternative, as care at the clinic was holistic. The midwives checked blood pressure and tested urine but also wanted to know about family dynamics and birthing dreams. Similarly, the homebirthing mothers I interviewed were familiar with the ideas of British obstetrician Grantly Dick-Read and French physician Fernand Lamaze; at the same time, many still have a cherished copy of *Spiritual Midwifery*, the seminal book by American midwife Ina May Gaskin, herself a back-to-the-land pioneer.[61] Perhaps reflecting the idiosyncratic way that counterculture connections worked, while Kootenay midwives were connected to homebirth activities outside the region, few of the birthing women I interviewed were aware of Cheryl Anderson's and Raven Lang's radical midwifery work in south coastal British Columbia or Lang's activism at the Santa Cruz Birth Center in California.

The key Kootenay childbirth educational text was *Responsible Home-Centred Childbirth: A Parents' Manual*, compiled in the late 1970s by the local Kootenay midwives. In the style of other back-to-the-land how-to manuals such as the *Whole Earth Catalog*, which also explained "nature" to the uninitiated, the ninety-two-page volume is comprehensive; it includes a detailed description of the processes of pregnancy, childbirth, and the post-partum period; information sections on nutrition, useful herbs, sexual relations, and what to do if the midwife did not arrive in time; and a list of eighty-nine books for

10.4 Cover of *Responsible Home-Centred Childbirth: A Parents' Manual*, compiled in the late 1970s by the local Kootenay midwives. Source: BC Midwifery Collection (Megan Davies, collector), University of British Columbia Archives, Vancouver.

Responsible Home-Centred Childbirth

A PARENTS' MANUAL

Kootenay Childbirth Counselling Centre

further reading.[62] Community midwife Barbara Ray told me that the purpose of the manual was to "take the mystique out of childbirth," thereby democratizing medical knowledge. With the entire process clearly laid out in the comprehensive volume, expectant mothers and their partners had to take responsibility for knowing "everything."[63] Parents were expected to be materially prepared for a homebirth with a sterile pack of sheets, washcloths, towels, rubbing alcohol, olive oil, Dettol, sterile water and gloves, herb teas for labour and delivery, a large bowl for the placenta, and a "good reliable vehicle with [an] extra gas reserve can."[64]

The process of labouring and giving birth beyond the biomedical gaze also offered powerful moments of experiential education and knowledge formation that bisected the spiritual and pragmatic realms. After Ellie Kremler's daughter Faith was born, midwife Abra Palumbo showed Kremler and her partner how the tree of life was visible in the placenta and encouraged them to eat the organ.[65] Similarly, Lisa Farr, a vegetarian for years at the time, believed that she was replenishing vital minerals lost during a post-birth hemorrhage when she took her midwife's advice and ate the cooked placenta.[66]

Community formation and the creation of social capital and individual and collective identity also operated on practical and pragmatic levels among the birthing families of the region, contributing to self-reliance as a collaborative rather than individual activity. Counterculture neighbours traditionally stocked up the freezer of a woman about to give birth.[67] In Argenta, the isolated former Quaker settlement on the east side of Kootenay Lake, a group of young families were "all doing it together" in the late 1970s. Neighbours shared childcare and took birth photographs for one another, and "everyone helped out after a birth." When Liz Tanner hemorrhaged after delivering her son Forest, the men of the community rescued her from the top floor of the family A-frame, sawing a hole through the bedroom floor, lowering Liz down on an old door, and driving mother and midwife to hospital in a station wagon.[68]

To a large extent, Kootenay counterculture homebirth and midwifery practice existed outside the late-twentieth-century economic system. Local community midwives knew that they would likely receive negligible financial remuneration for their work; as Ray told me, "It wasn't about the money."[69] In the early 1980s, the fee for a birth was normally one hundred dollars per midwife, but many cash-strapped parents paid fifty dollars—or nothing at all.[70] Rather, Kootenay residents frequently paid for midwifery services in kind, with natural products from the home environment, a reflection of the countercultural belief that privileged informal methods of economic exchange.[71] Ray accepted a year's worth of eggs, garden produce, and car maintenance in return for midwifery services.[72] Similarly, Bush remembered being compensated with homemade bread and canning, firewood, boxes of apples and garden produce, housecleaning, and childcare.[73]

Such articles and services reflect an understanding of homebirth as an individualized experience that was linked to older agrarian traditions and thus oppositional to an impersonal, biotechnological hospital birth. Some payments were specially handcrafted objects imbued with the meaning of the moment, like the teapot fashioned for Palumbo by potter and mother Pamela Stevenson, its lid the exact circumference of a fully dilated cervix.[74]

A similar counterculture interest in creating alternative identities through a celebration of nature and the handcrafted is reflected in the story of Kootenay birthing quilts, each block imbued with meaning and painstakingly produced by women. Here, nature and art serve both a personal purpose, as a unique and special gift for a tiny infant, and a social function, by fostering alternative culture and community. Examples of nostalgic rural female art, the images appliquéd and embroidered on the small squares frequently depict flowers and butterflies, both of which are emblematic of the idealized rural environment of the Kootenays. Some squares blend imagery from nature with deeper philosophical statements, like the tree of life depicted on the quilt made for midwife Ilene Bell's son Thomas.[75] Others, like the stork depositing an egg on Palumbo's daughter's quilt, reference the process of birth. Heart motifs that evoke the emotional space of homebirth, sometimes entwined with flowers, were also common. A wild rose detail from a 1984 birth quilt was a "message" about wild roses, suggested by an Aboriginal healer as an herbal remedy for that particular sick newborn baby.[76]

Collaborative crafts like quiltmaking are quintessential female activities and were strongly reminiscent of the imagined past that Kootenay back-to-the-landers were striving to recreate. Men might contribute a quilt block, but this was exceptionally rare. Typically, a group of women would work collectively to create the piece, each contributing a block that contained a personal message to mother and child. Quilts thus spelled out important messages about individuality, but within a collective context. Palumbo, who crafted a tiny suede fetus for a quilt in 1979, told me that the birth quilts reinforced the ethic that "You are not just born into a family, you are born into a

MEGAN J. DAVIES

10.5 Women and children inspecting baby Anna Palumbo's new birth quilt, in Perry Siding, BC, 1982. Source: Palumbo photo collection.

community."[77] The fact that each quilt was made by hand demonstrates the prioritization of homemade over purchased goods within the counterculture as well as the importance of collective process.

Historian Pamela Klassen observes that homebirthing women have the capacity to create spirituality at the same moment in which they are creating individual and collective identity.[78] Klassen's thoughtful analysis is useful in interpreting the Blessing Way, a procreation ceremony enacted by counterculture people of the Kootenays. A Blessing Way ritual could be a deeply intimate event, to bring peace after a miscarriage, or as large as the ceremony presented by the Kootenay Childbirth Counselling Centre as a workshop for mothers and babies at the 1984 Festival of Awareness, held at the David

10.6 Blessing Way ritual: Midwife Abra Palumbo bathing future mother Irme Mende's feet, in Lemon Creek, BC, 1981. Source: Palumbo photo collection.

Thompson University Centre and attended by an estimated seven hundred people.[79]

Barbara Ray explained that the Blessing Way honoured the act of birth as "not just a physical process, but a rite of passage, truly a transformation." The natural and the female were key motifs employed in the event. An appropriation and adaptation of a Navajo coming-of-age ritual, a Blessing Way usually took place before the birth of a child, but the changeable, individualistic nature of the event meant that it was also used to acknowledge a miscarriage or celebrate a marriage. Wearing colourful clothing and bringing a gift either handmade or infused with the special meaning of the moment, participants created a ring around the pregnant woman and her partner, both of whom were adorned in floral wreaths. The midwife, or sometimes the mother's closest female friends, washed her feet and massaged them with cornmeal.[80]

MEGAN J. DAVIES

The use of the Blessing Way in the Kootenays demonstrates local links both to radical American midwife Raven Lang, who was engaged in similar rites in California, and to the wider back-to-the-land interest in Aboriginal peoples and customs.[81] A handwritten description of the ceremony, created as a guide and carefully decorated with a flower motif, speaks to the purposeful creation of alternative birthing culture with a spiritual inspiration: "About 10 years ago, this ceremony sprang up in our culture, as a kind of new age baby shower, but with a more intimate and powerful focus on the birthing family. This concurred with the general trend to embrace and create traditions as a way of understanding and celebrating changes or passages and events in our lives."[82]

There is a clear connection between the creation of alternative collective and personal identities and the homebirth movement of the 1970s and 1980s in British Columbia's Kootenay region. However, this is a complex story, perhaps best understood through the lens of consumption. As scholars in the field have noted, maternalism has always positioned itself as oppositional to the world of the marketplace, represented in this context by medical expertise and the institutions of state and medicine.[83] Neo-traditional feminist—or rooted in a more diffuse set of alternative living practices—participants in back-to-the-land childbirth understood their quest to be about social relationships and creating new personal and collective ways of bringing the next generation into the world.

CONCLUSION

We were interested in taking more responsibility for ourselves and our lifestyle and how we were going to raise our children. How we were going to feed them. How we were going to live on the land. We had respect for the natural world and the primal nature of life. And giving birth is still a very primal process, rather than a cerebral one. It is not about doing it right or wrong.[84]

As Kootenay community midwife Barbara Ray makes apparent, the organic, social, and spatial environment was frequently evoked in homebirth narratives and constantly woven into aspects of the movement, from spiritual Blessing Ways and community birth quilts to childbirth education and the presentation of the female body in labour. Nature and place gave cohesion to this radical maternalist movement as it took shape and imparted personal and collective power to the practice of reclaiming birth in the Kootenays.

The story of counterculture homebirth in the Kootenays thus provides a case study of how back-to-the-land women embraced the living world, consciously seeking out a lifestyle mediated by nature and maternalism rather than technology and professionalism, and making real the feminist mantra of the era: "the personal is political." By shifting childbirth from the medical/professional realm of the hospital to the natural/personal sphere of the counterculture homespace, the birthing mothers I interviewed were able to assert ownership of their bodies and claim an alternative identity as self-educated, resourceful people who were part of a community of like-minded individuals.

Midwife-assisted homebirth in the 1970s and 1980s therefore needs to be understood as a complex process of decolonization, re-location, re-education, and rediscovery. In the long valleys of the Kootenays homebirthing women drew on a wide set of cultural and social identities, resources, and capital, rejecting a biomedical model of childbirth in favour of a holistic approach that incorporated the natural alongside the communal, the relational, the female, and the spiritual. The production of the natural and reproduction of family and community were thus one and the same. Although the descriptor "female labour" takes on multiple meanings here, this work simultaneously situated women within the traditional maternal role and allowed them to radically transcend the lived experiences of their mothers by reclaiming childbirth and motherhood as female, political, and empowering.

NOTES

1   This project would not have been possible without funding from Associated Medical Services (Toronto). I would like to thank the following individuals for their help with the project and this paper: Lanny Beckman, Michele Billung-Meyer, Lara Campbell, Colin Coates, Sarah Gose, Christopher Hines, Jude Kornelson, Sarah Jane Swartz, and all the women that I interviewed, especially Abra Palumbo for her ongoing engagement with the project.

2   Pamela Nagley Stevenson, videotaped interview with the author, Winlaw, BC, 28 June 2004.

3   Over the latter nineteenth century, physicians launched a successful bid to exclude midwives and position themselves as the only legitimate birthing experts. See Leslie Biggs, "Rethinking the History of Midwifery in Canada," in *Reconceiving Midwifery*, ed. Ivy Lyn Bourgeault, Cecila Benoit, and Robbie Davis-Floyd (Montreal: McGill-Queen's University Press, 2004), 17–47; and Mike Saks, "Medicine and the Counter Culture," in *Medicine in the Twentieth Century*, ed. Roger Cooter and John Pickstone (Amsterdam: Harwood, 2000), 113–23. On the shift to hospital birth and the rise of obstetrics, see Wendy Mitchinson, *Giving Birth in Canada, 1900–1950* (Toronto: University of Toronto Press, 2002).

4   For Canadian literature on these topics, see Brian Burtch, *Trials of Labour: The Re-emergence of Midwifery* (Montreal: McGill-Queen's University Press, 1994); Sheryl Nestel, *Obstructed Labour: Race and Gender in the Re-emergence of Midwifery* (Vancouver: UBC Press, 2006); and Margaret MacDonald, *At Work in the Field of Birth: Midwifery Narratives of Nature, Tradition and Home* (Nashville: Vanderbilt University Press, 2007). For a comprehensive picture of the American scene, see Margot Edwards and Mary Waldorf, *Reclaiming Birth: History and Heroines of American Childbirth Reform* (New York: Crossing Press, 1984). Eleanor Barrington's *Midwifery Is Catching* (Toronto: NC Press, 1984), a polemic designed to promote midwifery, is also an important historical document with an extensive section on the Kootenays.

5   The area's unconventional orientation was reflected in its birth statistics: a 1980 federal study found that 8 percent of Kootenay births took place at home, compared with 1 or 2 percent elsewhere in the province. Cited in Barrington, *Midwifery Is Catching*, 87. Ina May Gaskin and her cadre of midwives worked on The Farm in rural Tennessee. Their work has recently been chronicled in the documentary *Birth Story: Ina May Gaskin and the Farm Midwives*, directed by Sara Lamm and Mary Wigmore (Los

Angeles, 2012). For a history of The Farm, see Timothy Miller, *The 60s Communes: Hippies and Beyond* (Syracuse: Syracuse University Press, 1999), 118–24. For a history of Lang and the Santa Cruz Birthing Center, see Edwards and Waldorf, *Reclaiming Birth*, 156–79.

6 Because homebirth was located in the family setting, and appears to have been primarily a heterosexual scene in the Kootenays during this period, men were clearly part of the picture. But the key issue that emerged through my research was the development of a female-controlled women-centred birthing system. It is hoped that future research will explore the sexual and gender dynamics of the homebirth movement.

7 In 2000 I conducted eight preliminary interviews in the Kootenays, and in 2004 I conducted twenty-nine videotaped interviews with women who had worked as "community" midwives and/or given birth at home during the 1970s and 1980s in the Kootenays, the southern Gulf Islands, and Vancouver. These are in the University of British Columbia Archives, Vancouver (hereafter UBC Archives), along with Abra Palumbo's midwifery papers. The personal connections that I established with some of my research subjects preclude any pretense at objectivity. Complete scholarly detachment is impossible.

8 Theoretical works on social and cultural capital include Robert Putman, "Bowling Alone: America's Declining Social Capital," *Journal of Democracy*, 6, no. 1 (1995): 65–78; and Pierre Bourdieu, *Practical Reason* (Stanford: Stanford University Press, 1998).

9 Carolyn Merchant, "Reinventing Eden: Western Culture as a Recovery Narrative," in *Uncommon Ground: Rethinking the Human Place in Nature*, ed. William Cronon (New York: W. W. Norton, 1996), 132–59.

10 Gregg Mitman, "In Search of Health: Landscape and Disease in American Environmental History," *Environmental History*, 10, no. 2 (2005): 184–210.

11 See Burtch, *Trials of Labour*, 3–4. The legalities of homebirth were presented most strongly by Liz Tanner in recounting the hemorrhage that followed one homebirth. Diane Holt, Susan Stryck, Liz Tanner, and Don Tanner, videotaped interview with the author, Nelson, BC, 28 June 2004.

12 Aronsen neglects important social justice and radical health initiatives, but his book is a good introduction to the Vancouver scene—the centre of much of the BC action in the period. Lawrence Aronsen, *City of Love and Revolution: Vancouver in the Sixties* (Vancouver: New Star, 2011). For a national perspective, see Doug Owram, *Born at the Right Time: A History of the Baby Boom Generation*

(Toronto: University of Toronto Press, 1996); and Myrna Kostash, *Long Way from Home: The Story of the Sixties Generation in Canada* (Toronto: James Lorimer, 1980).

13   Justine Brown, *All Possible Worlds: Utopian Experiments in British Columbia* (Vancouver: New Star, 1995); on British Columbia's back-to-the-land movement, Terry Allan Simmons, "But We Must Cultivate Our Garden: Twentieth Century Pioneering in Rural British Columbia" (PhD diss., University of Minnesota, 1979); for studies dealing with other locales, Alan MacEachern, *The Institute of Man and Resources: An Environmental Fable* (Charlottetown, PEI: Island Studies, 2003); and Miller, *The 60s Communes.*

14   Mark Vonnegut, *The Eden Express* (New York; Praeger, 1976), 48. Other BC accounts of the back-to-the-land movement include Kenneth Brower, *The Starship and the Canoe* (New York: Bantam, 1978); and Feenie Ziner, *Within This Wilderness* (New York: W. W. Norton, 1978).

15   For a Canadian introduction to social movement theory, see Marie Hammond-Callaghan and Matthew Hayday, eds., *Mobilizations, Protests and Engagements: Canadian Perspectives on Social Movements* (Halifax, NS: Fernwood, 2008).

16   See the chapters by Janovicek and Rodgers in this collection.

17   On the American feminist health scene, see Morgan, *Into Our Own Hands.* Key feminist theorists from the period who wrote about women and birth include Ann Oakley, *Women Confined: Towards a Sociology of Childbirth* (Oxford: Martin Robertson, 1984); Adrienne Rich, *Of Woman Born: Motherhood as Experience and Institution* (London: Virago, 1977); Barbara Ehrenreich and Deidre English, *Witches, Midwives, and Nurses: A History of Women Healers* (New York: Feminist Press, 1973).

18   While naturist practices have a much longer Canadian history, public nudity was an important aspect of counterculture rebellion against mainstream morality. Linked to liberal sexual practices, dress, and hairstyles, nudity found local expression in a Nelson nudist commune of the late 1960s and in practices elsewhere such as nude sports events and the establishment of nude beaches. Lanny Beckman, personal communication, July 2012; John Davies, Brenda Davies, Elaine Davies, and Michael Davies, personal communication, June 2005; Aronsen, *City of Love,* 70.

19   The history of alternative medicine is still limited. See J. K. Crellin, R. R. Anderson, and J. T. H. Connor, eds., *Alternative Health Care in Canada: Nineteenth- and Twentieth-Century Perspectives* (Toronto: Canadian Scholar's Press, 1997); Mike Saks, "Political and Historical

Perspectives," in *Perspectives on Complementary and Alternative Medicine*, ed. Tom Heller, Geraldine Lee-Treweek, Jeanne Katz, Julie Stone, and Sue Spurr (Abington, UK: Routledge, 2005), 59–82; Saks, "Medicine and the Counter Culture."

20 Pat Armstrong's biographical details come from Barrington, *Midwifery Is Catching*, 87–88; and field notes, November 2000, Nelson, BC.

21 My thoughts on the rural context have been stimulated by discussions of the English countryside presented in Paul Milbourne, ed., *Revealing Rural "Others": Representation, Power and Identity in the British Countryside* (London: Pinter, 1997); and Keith Halfacree, "'Back to the Land'? (Re)settlement as a Future for the Next Millennium," in *The New Countryside: Geographic Perspectives on Rural Change*, ed. Kenneth B. Beesley, Hugh Millward, Brian Ilbery, and Lisa Harrington (Brandon, MB: Brandon University, 2003), 268–77.

22 Popular in the post–World War II era, but particularly in the 1960s, an A-frame house is designed in an inverted "V" shape, with a roof that extends to the ground.

23 Sarah Hardy and Caroline Wiedmer discuss these important and often contradictory geographies of the home and the ideologies that are at play in this space in "Introduction: Spaces of Motherhood," in *Motherhood and Space: Configurations of the Maternal through Politics, Home and the Body*, ed. Sarah Hardy and Caroline Wiedmer (New York: Palgrave Macmillan, 2005), 1–14.

24 Allison M. Williams's work on dying at home is most useful in theorizing homebirth in a locational context. Williams, "The Impact of Palliation on Familial Space: Home Space from the Perspective of Family Members Who Are Living (and Caring) for Dying Loved Ones at Home," in *Home/Bodies: Geographies of Self, Place, and Space*, ed. Wendy Schissel (Calgary: University of Calgary Press, 2006), 99–120.

25 Bell hooks argues that the "homeplace" is a space from which alternatives can be tested and resistance to dominant ways of seeing can be established, providing self-dignity for those involved; see bell hooks and Cornel West, *Breaking Bread: Insurgent Black Intellectual Life* (Toronto: Between the Lines, 1991).

26 Margaret MacDonald notes the importance of traditional historical images of midwives within the Canadian midwifery movement. MacDonald, "Tradition as a Political Symbol in the New Midwifery in Canada," in Bourgeault, Benoit, and Davis-Floyd, *Reconceiving Midwifery*, 46–66.

27 Stevenson, interview.

28 Bennett M. Berger makes this point, arguing that the presence of the father and other

children and the exclusion of the physician symbolized that the child belongs to the family, not the state. Berger, *The Survival of a Counterculture: Ideological Work and Everyday Life among Rural Communards* (Berkeley: University of California Press, 1981), 54.

29 Veronica Strong-Boag, "Home Dreams: Women and the Suburban Experiment in Canada, 1945–60," *Canadian Historical Review* 72, no. 4 (1991): 471–504.

30 Elaine Tyler May, *Homeward Bound: American Families in the Cold War Era* (New York: Basic, 1988), 172.

31 Leslie M. Campos, Lisa Farr, Irma Mende, and Lynnda Moore, videotaped interview with the author, Slocan, BC, 27 June 2004.

32 Kostash argues that "What they [the back-to-the-landers] tended to construct instead was a nostalgic throwback to the social organization of the frontier, where pre-industrialized society was speciously secured in the servitude of women" (*Long Way from Home*, 121).

33 Gretchen Lemke-Santangelo, *Daughters of Aquarius: Women of the Sixties Counterculture* (Lawrence: University Press of Kansas, 2009), 57–58, 59.

34 Laurel Robertson, Carol Flinders, and Bronwyn Godfrey, *Laurel's Kitchen: A Handbook for Vegetarian Cookery and Nutrition* (Tetluma, CA: Nilgiry, 1976); Wendy Bender, *Earth Market Cookbook: Recipes for a Simple Life* (Sooke, BC: Fireweed, 1978).

35 Warren Belasco, "Food and the Counterculture: A Story of Bread and Politics," in *The Cultural Politics of Food and Eating: A Reader*, ed. James L. Watson and Melissa L. Caldwell (Oxford: Blackwell, 2005), 217–34.

36 Lemke-Santangelo sets home birth in this context, seeing midwives as the ones who took "private" knowledge public (*Daughters of Aquarius*, 173).

37 Abra Palumbo, videotaped interview with the author, Black Creek, BC, 9 July 2004. See Megan J. Davies, "La renaissance des sages-femmes dans la région de Kootenay en Colombie-Britannique, 1970–1990," in *L'incontournable caste des femmes: Histoire des services de soins de santé au Québec et au Canada*, ed. Marie-Claude Thifault (Ottawa: University of Ottawa Press, 2012); and MacDonald, *At Work in the Field*, 131.

38 Saks, "Political and Historical Perspectives," 59–82.

39 Camille Bush, videotaped interview with the author, Vancouver, 13 July 2004.

40 Ibid.

41 Pat Armstrong, interview with the author, Nelson, BC, November 2000.

42 Birth albums, in the private collection of Carolyn deMarco.

43 Raven Lang, *The Birth Book* (Ben Lomond, CA: Genesis,

1972); Cheryl Anderson, *Free Delivery* (Vancouver: n.p., 1975); Ina May Gaskin, *Spiritual Midwifery* (Summertown, TN: Book Publishing Company, 1972). Anderson's book is a particularly good illustration of this point.

44   Marguerite S. Shaffer, "On the Environmental Nude," *Environmental History* 13, no. 1 (2008): 126–39.

45   Ilene Bell, article in *Maternal Health News* (Vancouver) 8, no. 3 (1983).

46   For an excellent discussion of maternalism and its connections with economic and political issues and cultures of female esteem and honour, see Rebecca Jo Plant and Marian van der Klein, "Introduction: A New Generation of Scholars on Maternalism," in *Maternalism Reconsidered: Motherhood, Welfare and Social Policy in the Twentieth Century*, ed. Marian van der Klein, Rebecca Jo Plant, Nichole Sanders, and Lori R. Weintrob (New York: Berghahn, 2012), 1–21.

47   Campos et al., interview.

48   Ellie Kremler, interview with the author, Kaslo, BC, 30 June 2004.

49   Campos et al., interview.

50   Saks identifies personal responsibility as a central aspect of the alternative health field in the late twentieth century ("Medicine and the Counter Culture," 113–23). Lemke-Santangelo places such themes as reconnecting with nature, self-realization, and spiritual growth in the broader context of the counterculture (*Daughters of Aquarius*, 56–57).

51   Arthur W. Frank, *The Wounded Storyteller: Body, Illness, and Ethics* (Chicago: University of Chicago Press, 1995).

52   Prenatal class handouts, personal papers of Abra Palumbo, UBC Archives.

53   Barrington, *Midwifery Is Catching*, 88.

54   Barbara Ray, videotaped interview with the author, Victoria, BC, 10 July 2004.

55   Glenda Patterson and Celestiana Hart, interview with the author, Nelson, BC, 2000. Pat Armstrong explicitly stated that *Our Bodies/Ourselves*, the seminal North American feminist health text, was not on her reading list. Armstrong, interview. As MacDonald points out in her work on Ontario midwives, they share the beliefs that birth should be a profound female life-event; midwifery care should be low-intervention, personalized, and woman-centred; and women should "give birth safely with power and dignity." MacDonald, "Tradition as a Political Symbol," 53.

56   Armstrong, interview.

57   Susan Vetrano, Liz Tanner, and Diane Holt, interview with the author, Nelson, BC, November 2000.

58   Morgan, *Into Our Own Hands*, 6. Other illustrations of the democraticization of expert

MEGAN J. DAVIES

knowledge in this period include the People's Law School in Vancouver and the establishment of crisis phone lines, hospice services, and so on.

59 Dr. deMarco came to the Kootenays in 1975; reading Raven Lang's *Birth Book* on her journey west, deMarco decided to attend homebirths and became a key mentor for the Kootenay midwives. Barrington, *Midwifery Is Catching*, 86–93. DeMarco is featured in the 1993 National Film Board documentary *Born at Home*.

60 Bush, interview.

61 Gaskin, *Spiritual Midwifery.*

62 *Responsible Home-Centred Childbirth* (Kootenay Childbirth Counselling Centre, n.d.). The manual was typed out by Abra Palumbo and illustrated by Patti Strom. Barbara Ray, videotaped interview with the author, Victoria, BC, 7 July 2004. For an analysis of how manuals worked in another counterculture community, see Joan Cross, "Capitalism and Its Discontents: Back-to-the-Lander and Freegan Foodways in Rural Oregon" *Food and Foodways*, 17, no. 2 (2009): 64.

63 Ray, interview, 7 July 2004.

64 *Responsible Home-Centred Childbirth*, unpaginated [at p. 65]. Diane Hold, Liz Tanner, and Susan Vetrano recalled with some amusement their efforts to sterilize linens (which they were meant to repeat every five days if the baby was late) inside a wood-burning stove—often with predictable results! Vetrano, Tanner, and Holt, interview.

65 Kremler, interview.

66 Lisa Farr also had Palumbo as a midwife. Campos et al., interview.

67 Barrington notes this in *Midwifery Is Catching*, 90.

68 Holt, Stryck, Tanner, and Tanner, interview.

69 Ray, interview, 10 July 2004. None of the Kootenay practitioners I interviewed made a living from their craft. Abra Palumbo was paid fifty dollars for a birth when she began in 1976, an amount that had increased to two hundred dollars by 1985. As a result, her family finances depended on her husband's teaching income. Palumbo, interview. Camille Bush also recalled getting fifty dollars for a birth. Bush, interview.

70 Ilene Bell, videotaped interview with the author, Nelson, BC, 29 June 2004.

71 Here I am referencing Belasco's work on counterculture food and the stress put on "natural" food prepared through traditional methods ("Food and the Counterculture," 221).

72 Ray, interview, 10 July 2004.

73 Bush, interview.

74 Palumbo, interview; Stevenson, interview. I drank tea from that pot when I visited Abra the following year.

75 Ilene Bell, email communication with the author, 10 April 2009.

76 Ibid.

77 Barrington, *Midwifery Is Catching*, 90; Palumbo, interview.

78 Pamela Klassen, "Procreating Women and Religion: The Politics of Spirituality, Healing, and Childbirth in America," in *Religion and Healing in America*, ed. Linda Barnes and Susan Sered (New York: Oxford University Press, 2003).

79 Siegfried Noetstaller, photographer, "Mothers and Babes Participated in a 'Blessingway Ceremony,'" *Nelson Daily News*, 29 March 1984, 6.

80 This description is gathered from the following sources: Ray, interview, 10 July 2004; Palumbo, interview; Stevenson, interview; Kremler, interview; "The Blessing Way Ceremony," handwritten and hand-decorated description, n.d., personal papers of Abra Palumbo, UBC Archives; Barrington, *Midwifery Is Catching*, 91.

81 Abra Palumbo thought that Patti Strom might have brought the idea of the Blessing Way to the Kootenays, but the connection she suggests with Raven Lang is also likely. Abra Palumbo, letter to the author, 2 June 2005. By the late 1970s, Lang was teaching the ritual to women attending the Institute for Feminine Arts in California. Edwards and Waldorf, *Reclaiming Birth*, 187–88. This is likely a case of counterculture adherents playing with something symbolically labeled "Indian," without having any real connection with Indigenous peoples, as Philip Deloria explores in his useful essay, "Counterculture Indians and the New Age," in *Imagine Nation: The American Counterculture of the 1960s and '70s*, ed. Peter Braunstein and Michael William Doyle (New York: Routledge, 2002), 159–88. Geographer Cole Harris, himself from this area, points out that Indigenous peoples of the Slocan Lake region were decimated by infectious disease brought by the colonizing powers and thus rendered invisible to future incomers, including his own family. Similarly, counterculture migrants of the 1970s and 1980s would likely have been unaware of an Indigenous presence in the area. Cole Harris, *The Resettlement of British Columbia: Essays on Colonialism and Geographical Change* (Vancouver: UBC Press, 1997), xvi.

82 "The Blessing Way Ceremony," UBC Archives.

83 Janelle S. Taylor, "Introduction," in *Consuming Motherhood*, ed. Janelle S. Taylor, Linda L. Layne, and Danielle F. Wozniak (New Brunswick, NJ: Rutgers University Press, 2004), 1–18.

84 Ray, interview, 10 July 2004.

MEGAN J. DAVIES

Children of the Hummus: Growing Up Back-to-the-Land on Prince Edward Island

Alan MacEachern, with Ryan O'Connor

The back-to-the-land movement of the 1970s was defined by choice. A segment of the North American population considered what contemporary civilization had to offer them and consciously rejected it. They abandoned their urban or suburban existence in favour of rural, out-of-the-way places. They renounced many of the trappings and traps of modernity and went in search of a life that would be simpler and more self-sufficient, that would bring them closer to nature, to their work, to their food, to their families. The back-to-the-land movement was in fact exceptionally deliberate as far as movements go because it demanded as its fundamental act voluntary relocation—that is, *moving*.[1]

And yet there was a subset of back-to-the-landers for whom there was no choice at all. These were the children who moved because their parents did or who were born into the lifestyle and so became back-to-the-land by default.[2] On the one hand, because of their limited or lack of other experiences, these children might be expected to have found the way of life more natural than their parents did. On the other hand, because most attended school and so integrated with the wider society on a daily basis, and because their lifestyle was not one of their choosing, they might be expected to have found the experience considerably more difficult than their parents did. Children of back-to-the-landers—ambassadors or double agents for both the movement and modern life simultaneously—were in that sense a real test of the movement and what it signified. Given that, it is surprising how fleetingly children appear in histories of the movement. Eleanor Agnew does not interview any children of back-to-the-landers in writing *Back from the Land*; Jeffrey Jacob spends only a few pages of *New Pioneers* discussing the work they performed. The cover photo of Dona L. Brown's *Back to the Land* shows a child in a garden, but children do not appear in the index.

In 2008 and 2009, Ryan O'Connor and I conducted a series of two dozen oral interviews with back-to-the-landers who had moved to Prince Edward Island in the 1970s, and we documented their experiences in an online exhibit.[3] PEI, on the east coast of Canada, is the nation's smallest province in terms of both size (5,700 square kilometres) and population (112,000 in 1971) and had been a popular destination for back-to-the-landers during the era. It had beautiful summers, arable soil, and—thanks to a century of rural depopulation—cheap, cheap land. It was small enough that one could, like Henry Thoreau at Walden, get away from it all while remaining secure in the knowledge that civilization, in some form, was never more than a few kilometres away. What's more, the island was developing an international reputation in this period for its interest in self-sufficiency and sustainability, as evident in the provincial government's establishment of the Institute of Man and Resources to encourage a transition to alternative energy and the federal-provincial funding of an experimental bioshelter, the Ark.[4] Back-to-the-landers came from across Canada and the United States and spread throughout PEI, with clusters developing around the Dixon Road area in the province's centre and the Iris and Gairloch-Selkirk roads areas in the east.[5] They took over rundown old farms or they built homes in forests that had never been cleared. Their arrival "from away," as islanders refer to the

ALAN MACEACHERN, WITH RYAN O'CONNOR

outside world, was both an assault on and a validation of the tranquil, agriculturally based society that had for the most part not yet left the land—one that seemed as close to the nineteenth century as to the twenty-first—so the new homesteaders were met with everything from open hostility to open arms. The back-to-the-landers whom O'Connor and I interviewed shared what it was like to enter PEI society, what the day-to-day nature of life and work entailed, and what led them either to give up the lifestyle or to continue it to this day.

Children figured heavily in these stories. Most strikingly, a number of those interviewed had been motivated to move back to the land to build a better life for and with their children, and yet it was ultimately children that returned many to a more conventional course. As Laurel Smyth stated, "Mostly after a certain amount of time everybody's kids and the demands of putting them into the school system and everything seemed to pull us all back from the land and into the interfacing with the business economy in order to make a living. . . . Hippies became bureaucrats. It was shocking for us all."[6] Such statements led O'Connor and me to realize that in focusing our attention on adults who had come to Prince Edward Island and failing to interview their children, we had missed not merely eyewitnesses of the back-to-the-land movement but some of its key actors.

As a second stage of this project, in 2011 and 2012, O'Connor and I interviewed eighteen people who had moved as children or been born into back-to-the-land Prince Edward Island households in the 1970s and early 1980s. Many of these were the children of those interviewed in our earlier project. They spoke about family dynamics, relations with other children, the greater back-to-the-land community, and how their upbringing shaped their later life. These interviews offer a distinctly different perspective of the movement. Most notably, whereas the older generation's varied stories of settling into PEI evoked the movement's diversity—as they arrived from all over the continent and all sorts of socioeconomic backgrounds, fleeing modernity, underemployment, or the draft, and seeking societal change or just a personal adventure—the children's stories of growing up emphasize the commonality of their experiences. The children also speak far less about the nitty-gritty details of living back-to-the-land and spend proportionally more time discussing their feelings about it.[7] Related to that, while both generations describe the lifestyle mainly in terms of anecdotes and warm memories, the children tell far fewer stories of exertion, poverty, and disillusionment. Whether because they remember the lifestyle through the gauze of childhood or because growing up in the lifestyle had made them inured to its hardships, the children of back-to-the-landers come across as, if anything, more loyal to the movement than those who had chosen it deliberately.[8]

WE WISH WE COULD MAKE THIS UP

"The Dixon Road was . . . my entire life," says Michael Stanley. His family moved to this well-wooded area in 1980, when he was two, and exploring the woods became his and his sisters' prime source of entertainment. His father was a potter, his mother a weaver, and the family also raised animals. As a child he helped his mother make cheese and yogurt and bread and pasta. The house, initially lacking electricity and water, was a hippie shack built of recycled barn board; his parents built onto it four times over the years, giving it something of a maze quality. It was filled with objects: books, art, and, in Stanley's words, basically anything that was homemade, sentimental, or just cool. Of his room he says, "It was pretty typical when I was younger. It was filled with posters of—" (he pauses) "—trees and leaves. Maybe that isn't so typical!" All in all, he states, "We had an all-encompassing little microcosm in the woods. I didn't know anything different until I had to go to school, and I realized, 'Ooh, I'm not like all the other kids around here.'"[9]

Those who moved back to the land at a later age knew straight away how their lifestyle was different. Matt Zimbel was fourteen and living what he calls a *Mad Men*–style suburban existence outside New York City when his parents visited PEI in 1970 and fell in love with it. The family bought a rundown farm and one hundred acres in Argyle Shore with plans to live there the following year. "And then we came up next year . . . and drove right by it. 'That's not our house, that's a

ALAN MACEACHERN, WITH RYAN O'CONNOR

. . . shack.' Grass was all overgrown, the winter was brutal, it had done more demolition to it, and we kind of turned around and said, 'Holy shit, that's our house.' I certainly thought we were on a family adventure." Zimbel's father was an accomplished commercial photographer, and the family had no farming experience, but they dived headlong into a back-to-the-land existence, raising animals and producing ever more of their own food. They made butter and jam, milked cows, grew vegetables, dug clams, fished, and had a smokehouse. (Zimbel recalls the children having a pet lamb named Joey, and then his mother being absent one night and the rest of the family eating lamb for dinner: "Me and my brothers and sisters said, 'Dad—is this Joey?' And Dad said yes. And we said, 'He's good.'") Zimbel can't say he was pleased to be torn out of New York as a teen, yet he found it all interesting. But when his mother started making shampoo, "I said to myself, 'That's it—she's gone too far. That's fuckin' crazy.'"[10]

Whether they were born into a back-to-the-land existence or had it thrust upon them, those interviewed certainly understand in retrospect that their childhood was unusual. And they understand its narrative possibilities.[11] As Zoe Morrison, who grew up on the Gairloch Road, states, "It's an easy way to be interesting . . . because I didn't have to choose it, or do it, really."[12] That Morrison is willing to diminish her claim to a lifestyle in which she was immersed—saying she did not "do" back-to-the-land—suggests how important this lack of choice is in how the children today evaluate their relationship to the movement. Their childhood is not fully their story—or it can be treated lightly, *as* a story—because it was determined for them by their parents. That is not unusual, of course. Chanda Pinsent was eleven when her family loaded their belongings into a VW van and left British Columbia in 1974, buying vacant land in a valley in South Granville and building a house that had no electricity or running water. Pinsent recalls her parents as always taking the time to explain their lifestyle decision—even when as a typically disaffected teen she would ask, "Why the heck are we doing this?"—and that the children respected that their parents had clear political, philosophical, and environmental motivations for "taking ourselves out of the system."[13]

The earliest stories that many tell are secondhand, given to them by their parents. Clea Ward, born shortly after her American parents had built a log cabin half a kilometre into the woods on the Selkirk Road, in 1975, was told of her bottle freezing in her crib at night.[14] Ahmon Katz, born months after his parents had left Kent State University in Ohio, in 1971, and so likely one of the first back-to-the-land children on PEI, was similarly told that he used to turn blue while crawling on the floor of their ramshackle house.[15] Laura Edell, born two years after her parents had moved from New York in 1972, was privileged with the bedroom next to the chimney, as it stayed somewhat warmer in winter.[16] Those interviewed told many stories such as these, focusing on the pioneering nature of their childhood. And these accounts are typically told with pride and delight that they had experienced such a life so late in the twentieth century. Ahmon Katz remembers his little brother Sam waking him at night so that Ahmon could stand vigil outside the outhouse; Sam remembers sitting on his boots in the two-seater to keep warm. The Katzes also say, "We wish we could make this up: we really did have to walk uphill one mile in the snow to the bus every morning"; the Gairloch Road, which they lived on, was not plowed in winter—but the payoff was sledding home in the afternoon.[17] The back-to-the-land children tend to list a catalogue of things they went without: electricity, indoor plumbing, telephones, televisions. But these are not offered as evidence of deprivation. Michael Stanley recognizes that for his parents the back-to-the-land movement must have been "baptism by fire. They wanted to do it, and they did a lot of it wrong. But us kids, we didn't notice the hardships."[18] Aryana Rousseau, whose parents moved from Montreal (where she is now a lawyer), had never even heard of the back-to-the-land movement until she was a teenager; the life she lived just seemed normal.[19]

Considering that *Little House on the Prairie* was one of the most popular television shows of the era, it was probably inevitable that its portrayal of nineteenth-century homesteading became a touchstone. Pinsent never liked it when guests compared her home to those on the show just because there was kerosene lighting, no plumbing, and

ALAN MACEACHERN, WITH RYAN O'CONNOR

lofts for children's bedrooms. "That used to drive me nuts!" she says. "'No, it's not!'" She would try to explain that what her family was doing was political, a rejection of capitalism.[20] It probably did not help that Pinsent had no way of watching the TV show the visitors were referencing, although she had read the books by Laura Ingalls Wilder. But Ward, who lived without electricity until age eight and without running water until high school, recalls loving the Little House series specifically because "it wasn't that far from my own experience." Her father would hook up a car battery to provide electricity for an hour of television each week, and it was *Little House on the Prairie* that the family would watch.[21] As a place to grow up, there were certainly differences between a little house on the nineteenth-century Prairies and one on twentieth-century PEI, but the similarities were sufficient to help some children of back-to-the-landers communicate their lives to peers—and perhaps to help them better understand their own lives, too.

ALL ABOUT WORK / WE WERE THERE TO PLAY

The Prince Edward Island back-to-the-landers interviewed in 2008 and 2009 were unanimous in describing how much work the life entailed. Gerald Sutton summarized the difference between back-to-the-landers and hippies by saying, "to go back-to-the-land you have to do the work."[22] The children interviewed this time spoke with far less unanimity about work—if they spoke about it at all. When growing up on the Gairloch Road, Zoe Morrison was given a list of chores every day, from cleaning out the root cellar to picking berries to eviscerating turkeys (her family raised turkeys, so her memories of childhood Christmases are tainted by wholesale turkey massacres).[23] Yet Morrison's best friend then and now, Clea Ward, herself in a back-to-the-land family, does not feel she worked any harder than did other island kids. In fact, she jokes that back-to-the-land parents lacked some of the authority to demand work that other parents held: "No one was ever going to beat us. And there was no TV to withhold."[24]

Even siblings had conflicting memories of childhood work. According to Chanda Pinsent, "It was all about work. . . . Our lives were defined by the work and the chores we did." The Pinsent children's first year on PEI was spent helping build the house. They also regularly cut, split, hauled, and stacked wood to heat the home; hauled milk cans of water down from the river; and fed, milked, and cleaned up after a menagerie of livestock, as well as performing seasonal duties such as haying and tending the market garden. But perhaps Chanda's experience was shaped by being the oldest child; her sister Celine remembers things quite differently. While Celine used to think she was given too much to do, she now believes that that was just a typical child's feeling, and that she and her siblings had about the same amount of work as other PEI farm children and likely less than those on high-production farms.[25]

For others, either there was not much work or it did not feel like work. Sam Katz talks with surprising fondness of being tasked with getting the family's water from the outdoor hand pump each morning: "I always thought of it as a fun thing. It was never a chore—the water was so fresh. 'Sammy, pour me a bucket.'"[26] Aryana Rousseau's mother loved gardening so much—a fact that undoubtedly contributed to her decision to leave Montreal to go back to the land—that the kids were never expected to do much.[27] Likewise, Vanessa Arnold, four years old in 1975 when her parents quit their Montreal teaching jobs and moved to PEI to apply the sustainable agriculture practices they had been reading about, calls herself just a "little helper" given manageable jobs like gathering eggs from the henhouse; growing up, she was in no way directed toward farming or gardening.[28] It may well be that many back-to-the-landers felt their children were making sufficient sacrifices just in living the life that had been chosen for them. Michael Stanley was free to get involved in any job that interested him, such as milking goats, but otherwise taking care of the garden or animals "was just something Mom and Dad did. We were there to play."[29]

Play figures heavily in back-to-the-land children's accounting of that time. They know that whether because their family was poor or

ALAN MACEACHERN, WITH RYAN O'CONNOR

self-sufficient, or both, they largely went without the toys that other kids had. "There was no Tickle-Me Elmo in my house," says Andy Reddin, much as Arnold, in a separate interview, declares, "There were no Barbies in my house." But, again, they do not offer this as evidence of deprivation. Reddin appreciates having been taught at a young age the value of simplicity and the perils of consumerism. And Arnold mentions Barbie only as a counterpoint to the simple pleasures she enjoyed growing up in what was to her one big playground; crawling through culverts was all the fun she could have possibly wanted.[30] Like Stanley extolling the woods of the Dixon Road, many of those interviewed describe childhood as a rural idyll. The confluence of back-to-the-land kids in the Gairloch and Selkirk roads areas "free-ranged" together in the woods, as Zoe Morrison puts it, playing kick-the-can, hide-and-seek, and cops-and-robbers.[31] The Katz brothers tell of playing "Discovery" in the woods with their mother: burying, finding, and occasionally losing forever old coins she had collected.[32] Aaron Koleszar, who also came of age on the Gairloch Road, would years later appear on the cover of *Time* as an anti-globalization activist being violently arrested during the 1999 anti–World Trade Organization "Battle of Seattle." To him, growing up back-to-the-land meant the freedom to explore the woods, trails, ponds, and beaches of Prince Edward Island. "I hadn't put words to this before this interview," he concludes, "but [it was] the feeling of freedom."[33] For the children of the movement, because they did not have the same work responsibilities as their parents, back-to-the-land may have, more than anything, meant back-to-nature.

WORLDS COLLIDE

It is ironic that Aaron Koleszar defines his back-to-the-land upbringing in terms of freedom, considering that he more than anyone interviewed was tyrannized at school because of his upbringing. Like many back-to-the-land children, Koleszar was singled out because of his hair, his clothes, and the food he brought for lunch. At first he was teased, but it got worse as the children got older and bigger. He was bullied regularly, and in grade nine, two boys picked a fight with him—one jumped on his head, giving him a concussion. Koleszar moved back to Toronto to live with his father.[34]

Whereas adult back-to-the-landers withdrew from the mainstream, they had their children bussed back into it every school day. Michael Stanley describes his first day of school as "One of the most traumatizing days of my life. . . . I wanted to stay in the woods with my mom and dad and the little life I had known." When the school bus arrived, he started crying, bit his mother on the hand, and ran off into the woods.[35] Clea Ward also tells of hiding from the school bus in the woods. She believed she had nothing in common with her classmates: "I felt like I was from another planet."[36] The back-to-the-land children were exotic creatures in rural PEI schools. Eryn Gibbs, who was born on PEI in 1975, speaks of being treated as an outsider because of his long hair, his corduroys, and even the fact that his family did not go to church; that his family, like many of the back-to-the-landers, was poor did not help matters.[37] "We were definitely different," says Morrison, stressing each word. "We were dirty, we were the kids with lice." When lice were found on her and her back-to-the-land friends, and the school was unable to reach their parents because they did not have phones, the children were made to sit on the steps of the school until the end of the day.[38]

The back-to-the-land children did what they could to fit in at school. Some banded together in the French immersion programs, in which they were overrepresented.[39] The younger brother of Chanda and Celine, Miles Pinsent was teased about his ponytail when he started grade one, so he cut if off.[40] Pan Wendt, whose family came to PEI in 1975, distinctly recalls having to learn his appropriate gender role, which had not been impressed upon him by his parents or their circle of friends.[41] And what could not be changed or learned could be faked. Home Economics class made Chanda Pinsent particularly anxious: "I didn't know how to turn on a stove because we had a wood stove. Or use a sewing machine, because my mother had a treadle sewing machine. And I didn't know how to use an iron because we had an old cast iron that stood on the top of the stove. I would cover that

ALAN MACEACHERN, WITH RYAN O'CONNOR

up. I was really good at watching what other people did and figuring it out, because I didn't want people to know we didn't have that stuff."[42] But some things could not be hidden so easily. In a province where the phone book was dominated by Macs and Mcs, the five K families in the Gairloch Road area—including the Katzes and the Koleszars— were conspicuous. At school, how could the Chandas, Keirans, Pans, and Ahmons help but stick out? At age nine, Pan Wendt decided that he wanted to be called "Steve Jones."[43]

School lunches were the single most mentioned element in the interviews, standing in for all the differences between the worlds of the back-to-the-land children and of mainstream PEI. Stanley begins his discussion of his school years by stating, "Hey, when you come to school with hummus, and everyone says you're eating cat barf—that's a way not to fit in at school!"[44] Hippie staples that had not yet hit the mainstream, such as hummus, yogurt, and alfalfa sprouts, were ripe for derision by classmates and had to be defended. Other foods simply could not be. When the Katzes told me of their raccoon sandwiches, I was as prepared as I could be, having heard about those famous sandwiches from others. I asked, Did you really take raccoon sandwiches to Belfast School? Absolutely, they insisted, because their father figured anything he killed was literally fair game. But it was mortifying, and Ahmon demonstrates how he would hunch over his meal, cocking his wrist to hide his sandwich while he ate it. It has been pointed out to him, he notes, laughing, that he still eats like that.[45] Many of the back-to-the-land children speak of having lusted after foods that bespoke normalcy: white bread and bologna sandwiches, in Celine Pinsent's case; Jos. Louis cakes in Ward's; bread and wieners in Koleszar's.[46] "Normal" food was the only "normal" thing that Vanessa Arnold ever remembers craving. She would fling her lunch under the greenhouse when she arrived home from school every day.[47] Many of the back-to-the-land children note the irony that the mainstream—and they themselves—have grown to prefer many of the foods they once disdained.

In some ways, states Celine Pinsent, the most daunting part of arriving on Prince Edward Island as a back-to-the-lander was "just trying to catch up, to figure out what these people were about." But the sisters feel that integration was ultimately made much simpler by the fact that farming played such a big part in island life. "We would have seemed weirder if we were on the edge of suburbia," Chanda notes. Even if their family's "Noah's Ark approach" to farming—two cows, two goats, etc.—was hardly the norm, their peers could and soon did relate to them as just other farm kids.[48] And if their family was poor, so were many other farm families. The Pinsents and other back-to-the-land children found common ground with island kids through shared activities, such as 4-H or school sports or music clubs, or simply by being kids. Arnold recalls almost with amazement a little girl—"She was definitely from the island"—who announced out of the blue that Arnold was her friend; they played outdoors all the time after that.[49] While a number of those interviewed believe they were not invited to some friends' homes because their family's lifestyle was associated with drugs, promiscuity, atheism, or God-knows-what-else, Celine Pinsent notes with amusement that she seemed to receive some invitations specifically so her friends' parents could grill her about how her family lived. While visiting a friend's home could be a treat—watching television is often cited—the stakes were higher when they visited yours, and the back-to-the-land children speak of being careful with whom they opened their home to. Stanley was originally nervous about having "worlds collide" when friends saw his back-to-the-land existence, but he eventually took their reactions as a way of measuring his friends.[50] Visitors could even help a child see her life in a new way; Celine Pinsent had considered the sleeping lofts in her house irritating because they lacked privacy, but she was delighted to find them a big hit when friends visited.[51]

Of all those interviewed, only Rosie Patch speaks—with refreshing tartness—of an utter refusal to integrate with her island counterparts. Calling herself "brainwashed in a good way" by back-to-the-land life, she states that although she knew she lacked some things that others had, "I consoled myself with the belief that I was more moral, that I was living the right way, that I would end up smarter than those other people who had Fruit Roll-Ups in their lunchbox." (Having said that,

ALAN MACEACHERN, WITH RYAN O'CONNOR

she later describes trading her mother's homemade fruit leather for her classmates' commercial variety.) Patch made few friends in high school, believing that "Those people really, really weren't as good as us.... Those people had no moral principles! They were not struggling for any causes. I got on a train and went across the country to protest against the logging of Clayoquot Sound, I started an environmental club in my high school, I was damned if I was going to be friends with these layabouts!"[52]

While most of those interviewed agree that back-to-the-land children were picked on more than other kids, the boys in particular, they also believe that relations improved over time. Zoe Morrison sees herself as being on the "front lines" of back-to-the-land children who went to her school, and that even her younger sisters' experiences were far different, in part because she and her peers had made the lifestyle more familiar. "And we were, of course, very cool."[53] Back-to-the-land children often introduced progressive and cosmopolitan ideas to their island peers; when the Katz brothers returned to PEI after two years in Florida and Detroit, "we brought breakdancing back to Belfast."[54] Aryana Rousseau, born just seven years after Morrison, agrees that things were easier for younger kids like her. "The road had been paved," she says, using a phrase that nicely captures how rural PEI was becoming suffused with more contemporary ideas, in part thanks to the influx of back-of-the-landers.[55] In retrospect, it is easy enough to say that the back-to-the-land children had never been all that different from their island peers anyway—they were overwhelmingly white, roughly middle-class Judeo-Christians, if not Anglo-Saxons— so their integration experience was surely more straightforward than, for instance, that of subsequent Asian and African immigrant groups. But such a statement is made from a twenty-first-century perspective of greater societal respect for diversity. The children of back-to-the-landers experienced a distance that had to be crossed, and if the crossing seems simple today, it is because they crossed it.

COMMUNITY

Back-to-the-land children may well have integrated more, or at least faster, with the broader PEI society than their parents did, but it was not as if their parents had no network. "My parents and [their] friends didn't have to integrate on a social level" with islanders, Rousseau maintains. "They had each other."[56] Although going back to the land would seem the consummate expression of self-sufficiency, the movement enlisted people with similar backgrounds and philosophies and gave them similar day-to-day challenges. They even lived close to one another because of a shared need for inexpensive real estate. The result was individualists who gravitated naturally to one another. And as the parents gravitated, so did the children.

"If there's one thing I hope that you take away," says Clea Ward, "it is this notion of community."[57] Like many of those interviewed, she considers the back-to-the-landers as a close-knit, extended family, replacing the ones their families had given up to come to Prince Edward Island and standing in for the networks that other islanders enjoyed.[58] Although a variety of interweaving networks of come-from-aways, artistic types, and left-leaning folks existed on PEI in this era, the back-to-the-land network is remembered as being a particularly close-knit one. Monica Lacey recalls that she and her sister were jealous that although her father was an artist who had built their home on the Appin Road, because her family did not grow their own food or go off the grid, they were not considered full members of the homesteading fraternity.[59] It is worth adding that although the PEI back-to-the-landers were part of a continental movement sharing inspirations and cultural products—the *Whole Earth Catalog*, Helen and Scott Nearing, the *Dignam Land* newsletter, etc.—there was no sense of community between those on the island and those elsewhere. A number of the back-to-the-land children mention not being aware of the movement's existence until they were adults; Ward, who studied law in Toronto and now works in New Brunswick, says she still has yet to meet a back-to-the-lander from beyond Prince Edward Island.[60]

ALAN MACEACHERN, WITH RYAN O'CONNOR

Many of the back-to-the-land children played together all the time growing up, particularly those who lived in the enclaves in the Dixon Road, Selkirk-Gairloch roads, or Iris areas. Neighbouring families had dinners and parties together and travelled to gatherings of sixty or more to celebrate major holidays or welcome the summer or winter solstice. The families gathered for building bees, to tap maple trees, to sculpt, to watch movies projected on a wall, or to pick chanterelles or fiddleheads.[61] To Ward, this closeness was particularly valuable in giving back-to-the-land children access to a whole community of role models. For example, many of the women in the movement were strong feminists, and so not only helped bring feminism to PEI, but also brought it directly to girls like her.[62] In similar fashion, Michael Stanley says that the movement taught him to interact comfortably with adults from a young age, to have conversations with people decades older than him.[63] Of course, from a child's perspective, being watched over by a group of available adults could also have its downside; Ward and Morrison tell of knocking down a sign at a barn dance and then having ten adults on ten occasions tell them what they must do to make things right.[64]

Chanda Pinsent, born in 1963 and one of the oldest back-to-the-land children in 1970s Prince Edward Island, offers a more critical perspective of the community, even while calling her upbringing "an overwhelmingly positive experience." She observed the sort of groupthink absurdities that can thrive within any small society. Why did so many of the back-to-the-landers, many of them without cars, choose to build at the end of very long lanes, making getting out in the long island winter that much more difficult? Why, for that matter, did back-to-the-land parents such as hers think it was cool to wear their rubber boots to town, their jeans jauntily tucked out? "That was one of the things that really bugged me," Pinsent recalls. "*Nobody* else did." Moreover, she saw with clear eyes the limits of the back-to-the-landers' progressive principles. Differences in body strength meant that the homestead lifestyle often encouraged women and men to fall into traditionally gendered work roles.[65] Having become friends with many of the young women—some of them still teens—who moved to

PEI with considerably older partners, Pinsent believes that it was an especially hard life for them, because so many of the domestic jobs that became their responsibility were made much more difficult without electricity or running water.[66] And yet the back-to-the-landers' social broadmindedness could create its own tempests. While Laura Edell feels that incestuous is too strong a word to describe the PEI back-to-the-land community, she does note that partner-swapping was relatively common.[67] As one back-to-the-lander wryly put it to me, there were more divorces than orgasms.

That the back-to-the-land community was close-knit did not preclude its children from also interacting with Prince Edward Islanders outside of school. A number of those interviewed speak fondly of the relationships they built with mainstream neighbours, much as adults interviewed earlier for the *Back to the Island* project had done. Ward tells of locals such as Margaret and Barney Doherty taking her family under their wing—helping them, for example, with the horses that they had bought without having any knowledge of how to raise or work them.[68] Similarly, Stanley recalls local farmer Alec MacDonald driving down the Dixon Road periodically just to make sure that the tenderfoot farmers were doing all right.[69] Matt Zimbel's principal memory of his time on PEI is of quitting school at age fifteen and taking a job on Eddie MacPhail's pig, cattle, and potato farm. "I didn't know anything about farming, but I was strong and I was available at 7:00 a.m. Monday to Friday. In Argyle Shore, that means you've got a lot going for you." He spent twelve-hour days working alongside long-time island farmers and grew particularly close to MacPhail and his wife. When years later the band that Zimbel formed, Manteca, performed at the Confederation Centre of the Arts, the very shy MacPhail came up afterward and hugged him—a moment Zimbel sees as the capstone of "a wonderful experience."[70] Of course, not all interactions with islanders were so affirmative, and the back-to-the-land children certainly realized that many of the negative opinions of their lifestyle expressed by PEI children were ones learned from PEI parents. But the preponderance of reminiscences about settling into the island scene are positive. Celine Pinsent argues that islanders had

ALAN MACEACHERN, WITH RYAN O'CONNOR

been successfully integrating people for generations, not by changing them but by accommodating them.[71] In this way, back-to-the-landers were just another wave of migrants who embraced their prior identity in a new home, even as that home meant their identity evolved—and the identity of the new home evolved in turn.

THE BUOY

Rosie Patch's family eventually gave up the back-to-the-land life-style. When Patch grew up and moved to Montreal, she missed her mother's homemade bread and phoned for the recipe. Her mother all but refused to tell her, saying that Patch lived in a big, cosmopolitan city and should just go buy bread. But Patch wanted to raise her own children with some of the same principles with which she had been raised, such as eating homemade food and not watching television. "My mom says, 'They've got to watch TV, or they'll be out of touch with their peers.' Out of touch with their peers!" she says with disbelief. Patch now sells bread out of her Montreal home for five dollars a loaf.[72]

Since most of the back-to-the-land parents either came to Prince Edward Island with children in tow or had children soon thereafter, they seldom had much of a head start on adjusting to their new life-style. "My mom always said that we [she and I] grew up together," says Morrison.[73] Ahmon and Sam Katz's father was from the Bronx and found rural PEI's nighttime quiet so alien that he slept with an axe under the bed.[74] Ward's parents were the urban children of professors and, she says, had never built anything in their lives before moving to Prince Edward Island and erecting a log cabin. She speaks not just of her parents but of the entire movement when she states, "They couldn't imagine that anything bad would happen to them, because nothing bad had ever happened to them."[75] For many, going back to the land was never meant to be more than a temporary experiment, or adventure. It is little wonder that almost all of the families living the life on Prince Edward Island in the 1970s were, to varying degrees and at different speeds, eventually drawn back into the mainstream. They acquired running water and electricity, a car and television. They took employment outside the home to augment and then replace their attempts at farming, gardening, and self-sufficiency. They moved off their homesteads or off Prince Edward Island altogether. The children figured heavily in these decisions, because they relentlessly connected the family to the outside world and because their parents worried about imposing their chosen lifestyle on them.[76] While some of the back-to-the-land children interviewed here have retained or returned to vestiges of their childhood lifestyle, not a single one lived back-to-the-land straight through to adulthood.

And yet all of those interviewed speak of their unusual upbringing as absolutely formative. To be sure, some speak of the movement and their part in it in highly romantic terms. Laura Edell, almost forty and working in advertising in Toronto, says she still sometimes dreams of "chucking it all and moving back in a VW bug." But what are we to make of this statement, and of the fact that she is proud of having lived back-to-the-land, knowing that she and her mother had moved off the farm by the time Edell was four?[77] Back-to-the-land is for many a source of stories of an otherworldly yet somehow more authentic existence. "I can tell my kids," says Vanessa Arnold from her chiropractic/pilates studio in Toronto, "that I was attacked by the rooster and I ran out screaming with a bleeding leg. What four- or five-year-old can have that experience and then pass that on to their kids?"[78]

But the influences clearly run deeper than memorable anecdotes. Andy Reddin talks, as many do, about how growing up in the woods gave him an appreciation for nature. Now twenty-seven, and having earned a master's degree in physics and travelled around Asia, Reddin has returned to PEI and is going back to the land, at least for a while, in a cabin he is fixing up on his parents' Bonshaw property.[79] Michael Stanley stayed on the island and became a potter. He notes the resurgence in the past few years of "next-generationers"—grown back-to-the-land children—who are once again buying land on the Dixon Road and once again embracing the lifestyle, at least to some extent.[80] The Katz brothers even argue that being born into the way

ALAN MACEACHERN, WITH RYAN O'CONNOR

of life made them adept at technological problem-solving, and thus actually better suited to the life than their parents were. Ahmon lives just up the hill from where he grew up, working as an artist, carpenter, and builder; Sam is a key grip in the New York City film industry.[81]

Many of those interviewed speak of their back-to-the-land upbringing more generally as a source of strength. Aryana Rousseau credits her childhood in Mount Vernon as, paradoxically, what allowed her to become a lawyer in a big firm in Montreal. "I had seventeen years living in the woods, in a safe environment with a wonderful family, and that prepared me to go out and see the world and see what else I could do." And like many interviewed, Rousseau draws strength in part from the conviction that she could return to her old life if she had to: "I know there's something else, something more real—I have a backup plan."[82] Some of such sentiment is surely just daydreaming—and built on the questionable premise that being a child in a back-to-the-land household prepares a person sufficiently to adopt it herself decades later. But it is also an expression of how growing up in an alternative lifestyle revealed to those interviewed the possibility of alternatives. Ward, a lawyer who now works as a career counsellor in a law school, is not sure if she will ever move back to the land (though she and her husband talk about it), but if her children wanted to, she would encourage them to go for it. "Nothing bad can come of it. You can always go back to the grid. The grid will be there."[83]

The Pinsent sisters provide an interesting reflection on growing up back-to-the-land on Prince Edward Island, if only because locally their family became, to some extent, poster children for asceticism. On the one hand, the Pinsents lived one of the most hard-core back-to-the-land existences about as far into the island's interior as it was possible to go; on the other hand, their father grew increasingly well-known through his long-time involvement with the provincial government's Small Farms Program. Chanda, who would receive a Commonwealth Scholarship, travel the world, and settle in New Zealand, says that what made her "overwhelmingly positive" childhood distinct was the degree to which family existence was concentrated around a single project, the farm: "It was always the farm, the farm, the farm." But her father sold the farm years after the children had left, and when Chanda was interviewed in 2011 she did not think she would even bother taking her youngest son to see it when she returned to PEI that summer. "It wasn't the land," she realizes now. "It wasn't the place, it was the doing it, and that's where my memories are. And once he sold it . . . Initially, I thought the tie was to the land, but now I don't think it is, it's to the experience."[84] Chanda's sister Celine received her doctorate in clinical psychology and now runs a management consulting firm in Quebec. She sees her back-to-the-land upbringing as having taught her that the world is a place worth exploring and as having given her the confidence to do so. She likens life to swimming in the ocean, in which back-to-the-land is a buoy she can always swim back to if she needs to.

And yet it is Celine Pinsent, of all of those interviewed, who most directly questions whether it was growing up back-to-the-land that really shaped her upbringing, or whether it was growing up rural or on Prince Edward Island that was key.[85] I myself was raised in a never-left-the-land household on PEI, our family having farmed the same ninety acres since the 1830s. We lived more of a pre-modern existence than most, but in the 1970s we were still working with horses, still milking by hand, and had just recently acquired indoor plumbing. When I hear Eryn Gibbs describe the novelty of living on an unpaved road or of using an outhouse, I think, "That's my childhood, too"—at least until he describes keeping goats in the house in the winter.[86] It is interesting to hear Eryn's sister Keiran, now attending law school in Montreal, say that growing up back-to-the-land taught her to value community, natural beauty, and stewardship of the land—all values I equate with my own childhood—but that she is unlikely to return to PEI because she never felt like an islander.[87] There is likely no way to pull apart the strands of influence, to determine to what degree it was living back-to-the-land that shaped the childhood of those interviewed—particularly because they were interviewed specifically because they had grown up back-to-the-land. What is striking is how few even attempt this determination, happy instead to identify their childhood first and foremost with back-to-the-land.

Zoe Morrison concedes that growing up back-to-the-land may have been akin to growing up in any rural Prince Edward Island household in the 1970s "but with really liberal parents." Whether in terms of environmentalism, gay marriage, or organic food, PEI and North America at large have in the interim adopted many of the counterculture's ideas, or else the counterculture folks were just somewhat ahead of their time. In turn, the back-to-the-land children interviewed are much more immersed in the mainstream than their parents were then—although not really any more than their parents are now. A distinguishing characteristic of the island's back-to-the-land community, Morrison recalls, was beach nudity. Her mother was, in the words of a friend, "militantly naked," and Morrison herself did not own a bathing suit for part of her childhood. But at the back-to-the-land reunion in 2012 where some of these interviews took place, families met at a local beach, and Morrison was amused to see that islanders and back-to-the-landers alike were not only dressed, but dressed alike. You could not tell them apart.[88]

Looking at the back-to-the-land movement through the eyes of its children offers insight into how this process occurred. How has North American culture absorbed the counterculture to the point that they now resemble one another? Most accounts and analyses of the movement are from the perspective of its adult practitioners, who explicitly rejected the mainstream and so tended to see it as backsliding if they gave up the lifestyle or as an affectation if the broader society adopted elements of it. But for their children, rejection of mainstream society was never an option, in the sense that school immersed them in it every day and that their way of life was not of their choosing anyway. (One might argue that this lack of agency, and so accountability, actually makes the children more reliable assessors of the lifestyle than their parents are.) The back-to-the-land children interviewed here escaped some of the worst features of the lifestyle—the backbreaking labour, the awareness of deprivation—and instead experienced it in terms of family, play, and nature. They then shared their lifestyle—one involving environmental consciousness, self-sufficiency, simplicity, and hummus—with their peers, which helped permeate the broader culture even as the broader culture was permeating theirs. It is not that the back-to-the-land children turned their backs to the land while their parents toiled on it, but rather that they were by necessity always looking outward, facing the wider world.

NOTES

1   On the American movement of the 1970s, see Dona L. Brown, *Back to the Land: The Enduring Dream of Self-Sufficiency in Modern America* (Madison: University of Wisconsin Press, 2011), esp. chap. 7; Eleanor Agnew, *Back from the Land: How Young Americans Went to Nature in the 1970s and Why They Came Back* (Chicago: Ivan R. Dee, 2004); and Jeffrey Jacob, *New Pioneers: The Back-to-the-Land Movement and the Search for a Sustainable Future* (University Park: Pennsylvania State University Press, 1997). On the Canadian movement, see Sharon Weaver, "First Encounters: 1970s Back-to-the-Land Cape Breton, Nova Scotia and Denman, Hornby and Lasqueti Islands, British Columbia," *Oral History Forum d'histoire orale* 30 (2010): 1–30; Sharon Weaver, "Coming from Away: The Back-to-the-Land Movement in Cape Breton in the 1970s" (MA thesis, University of New Brunswick, 2004); Amish C. Morrell, "Imagining the Real: Theorizing Cultural Production and Social Difference in the Cape Breton Back-to-the-Land Community" (MA thesis, Ontario Institute for Studies in Education, University

of Toronto, 1999); and Terry Simmons, "But We Must Cultivate Our Garden: Twentieth Century Pioneering in Rural British Columbia" (PhD diss., University of Minnesota, 1979). It is difficult to calculate the scale of the movement; Simmons's estimate (pp. 192–94) that roughly one million North Americans moved back to the land in this period has become common in the literature (see, for example, Jacob, *New Pioneers*, 3; and Agnew, *Back from the Land*, 5) but, more recently, has been challenged (Brown, *Back to the Land*, 206).

2   There is no entirely satisfactory way to designate these children. "Children of back-to-the-landers" diminishes their participation in the lifestyle, while "children back-to-the-landers" or "back-to-the-land children" overstates how much agency they actually had—and, particularly for those born into the movement, raises the question of how they can be said to have moved "back."

3   The exhibit consists of a narrative history, a series of photographs by George Zimbel, and the audio recordings of

the interviews themselves. Alan MacEachern and Ryan O'Connor, *Back to the Island: The Back-to-the-Land Movement on PEI*, on NiCHE website, accessed 24 September 2015, http://niche-canada.org/member-projects/backtotheisland/home.html.

4   Alan MacEachern, *The Institute of Man and Resources: An Environmental Fable* (Charlottetown, PEI: Island Studies, 2003). See also Henry Trim's chapter, "An Ark for the Future," in this volume.

5   There were several short-lived communes on the island, but the vast majority of back-to-the-landers homesteaded as families, couples, or individuals. Although a couple of those whom O'Connor and I interviewed in our two projects had spent time in communes, all spoke exclusively of back-to-the-land homesteading experiences. As Brown notes in *Back to the Land,* 275n11, the commune movement peaked around 1970, when the homesteading movement was just catching on.

6   The *Back to the Island* interviews, plus brief biographies of those interviewed, can be found on the NiCHE website, http://niche-canada.org/member-projects/backtotheisland/interviews.html.

7   Which is not to say, of course, that children of back-to-the-landers provide an emotionally deeper or "truer" portrait of the movement than their parents

do. All oral history must deal with the reality that, as historian of childhood Neil Sutherland puts it, "[A] memory is really a *reconstruction* of what is being recalled rather than a *reproduction* of it." This is all the more the case when adults are being asked to recall their childhood—a time deeper in the past, when their personality was still being developed. And yet, as Sutherland writes, "if we are ever to get 'inside' childhood experiences, then we must ask adults to recall how they thought, felt, and experienced their growing up." Neil Sutherland, "When You Listen to the Winds of Childhood, How Much Can You Believe?" *Curriculum Inquiry* 22, no. 3 (1992): 239, 243. Sutherland's essay is an excellent introduction to issues surrounding the use of oral history of adults to recreate the internal worlds of children; it has informed our thinking here.

8   It must be said that in both series of interviews those who were more content with their back-to-the-land experience were surely overrepresented, since they were more likely to still be living the life, to be attending the 2012 reunion where some of the interviews occurred, or simply to be willing to talk about that time in their lives.

9   Michael Stanley, interview by Ryan O'Connor, 16 June 2011, 0:10, 36:30. Times given are for the beginning of new sections,

as seen in the interview transcripts.

10 Matt Zimbel, interview by Ryan O'Connor, 1 June 2011, 0:10, 7:00, 9:15.

11 Of moving from Manhattan to PEI, Matt Zimbel states, "My mother said, 'Children, one island's the same as the next, get in the goddamn car.'—That's not true, she never said that, but I think it's a good story." Ibid., 4:45.

12 Zoe Morrison, interview by Alan MacEachern, 4 August 2012, 29:30.

13 Chanda Pinsent interview by Ryan O'Connor, 15 May 2011, 16:15. The experience of Chanda and her sister Celine was distinct from that of all others interviewed here, in that they had been living in a commune before coming to PEI; in some ways, they found the family-focused nature of homesteading positively conventional by comparison.

14 Clea Ward, interview by Alan MacEachern, 4 August 2012, 40:00.

15 Ahmon Katz and Sam Katz, interview by Alan MacEachern, 4 August 2012, 0:10. My interview with the Katz brothers hardly satisfied oral history best practices. The only time and place we could arrange an interview was in pitch darkness in a tepee during the Home Again reunion, as a band played enthusiastically in the background. The fact that the two brothers were so often in agreement, to the point of ending each other's sentences, makes it difficult to determine which brother is speaking in the recording—even as it also makes such a determination somewhat less critical.

16 Laura Edell, interview by Alan MacEachern, 4 August 2012, 1:00.

17 Katz and Katz, interview, 5:50, 8:30.

18 Stanley, interview, 7:08.

19 Aryana Rousseau, interview by Ryan O'Connor, 5 June 2011, 1:20.

20 Chanda Pinsent, interview, 9:30.

21 Morrison, interview, 8:50, 12:40; Home Again roundtable, moderated by Ryan O'Connor, 5 August 2012, 20:45.

22 Gerald Sutton, interview by Ryan O'Connor, 13 June 2008.

23 Morrison, interview, 14:15; Home Again roundtable, 50:30.

24 Ward, interview, 10:30; Home Again roundtable, 47:00.

25 Chanda Pinsent, interview, 3:45; Celine Pinsent, interview by Ryan O'Connor, 20 April 2011, 44:00.

26 Katz and Katz, interview, 4:50.

27 Rousseau, interview, 2:50.

28 Vanessa Arnold, interview by Ryan O'Connor, 5 May 2011, 0:00.

29 Stanley, interview, 15:50.

30 Andy Reddin, interview by Alan MacEachern, 4 August 2012, 5:45; Arnold, interview, 23:10.

31 Morrison, interview, 17:00. The Katz brothers also describe such play. Katz and Katz, interview, 12:00.

32 Katz and Katz, interview, 27:20.

33 Aaron Koleszar, interview by Ryan O'Connor, 29 May 2011, 32:40.

34 Ibid., 5:45.

35 Stanley, interview, 12:10.

36 Clea Ward, comments, Home Again roundtable, 19:00. As Laura Edell drily notes, the back-to-the-land and local youth had only pot dealers in common. Edell, interview, 4:30.

37 Eryn Gibbs, interview by Ryan O'Connor, 14 April 2011, 6:50, 12:40.

38 Morrison, interview, 4:55.

39 French immersion arrived on PEI at about the same time as the back-to-the-landers, in 1975. From our standpoint, it may seem that these back-to-the-land parents were making a rather bourgeois, integrationist choice on behalf of their children, but it is as likely that, just as they expressed difference from the dominant culture in their way of life, they sought to open their children to different languages and cultures. On the prevalence of back-to-the-land children in French immersion programs, see Rousseau, interview, 6:50; Keiran Gibbs, interview by Ryan O'Connor, 28 April 2011, 2:15; and Ward, interview, 14:00.

40 Chanda Pinsent, interview, 7:50.

41 Pan Wendt, interview by Ryan O'Connor, 28 April 2011, 29:30. Wendt's story was unusual here, in that his family was part of a small community of "Premies" who came to Prince Edward Island in the 1970s to follow the religious and meditative teachings of the Divine Light Mission, led by Guru Maharaji Ji (Prem Rawat).

42 Chanda Pinsent, interview, 19:30.

43 Pan Wendt, comments, Home Again roundtable, 84:00.

44 Stanley, interview, 12:10. On other mocked foods, see also Celine Pinsent, interview, 38:30.

45 Katz and Katz, interview, 14:45. In another vein, Pan Wendt had to constantly explain his vegetarianism to peers—and explain it once again to their mothers at birthday parties. Wendt, interview, 25:00.

46 Celine Pinsent, interview, 38:30; Ward, interview, 46:00; Koleszar, interview, 7:10.

47 Arnold, interview, 21:00.

48 Chanda Pinsent, interview, 11:00; Celine Pinsent, interview, 4:10, 8:50.

49 Arnold, interview, 5:00.

50 Stanley, interview, 34:00. See also Home Again roundtable, 28:00.

51 Celine Pinsent, interview, 38:30

52 Rosie Patch, comments, Home Again roundtable, 15:30, 30:00.

53 Morrison, interview, 4:55, 29:00.

54 Katz and Katz, interview, 9:40.

55 Rousseau, interview, 4:50.

56 Ibid., 7:50. More than this, Clea Ward suggests that some back-to-the-landers were not particularly tolerant and had no interest in becoming friends with islanders. Home Again roundtable, 94:30.

57 Ward, interview, 48:30.

58 On this theme, see Morrison, interview, 4:30, 9:00; Rousseau, interview, 7:50, 13:20; and K. Gibbs, interview, 2:15. The back-to-the-land children's own grandparents and extended family did appear in their lives, of course. Several interviewees touch on how much difficulty their grandparents had with their parents' decision. According to Michael Stanley, his grandmother demanded the family buy a flush toilet because she was convinced he would be traumatized by going to school without knowing how to use one. He thinks it more likely she did not like using the outhouse when she visited. Stanley, interview, 10:50.

59 Monica Lacey, interview by Ryan O'Connor, 28 May 2011, 6:10.

60 Ward, interview, 40:00.

61 Back-to-the-land community social activities are discussed in Ward, interview, 26:40; Stanley, interview, 17:20; Edell, interview, 14:10; and E. Gibbs, interview, 3:00.

62 Ward, interview, 26:40, 29:30.

63 Stanley, interview, 17:20.

64 Clea Ward and Zoe Morrison, comments, Home Again roundtable, 10:00.

65 For example, Eryn Gibbs states that although his parents tried to assign chores indiscriminately to him and his sisters, he ended up doing most of the outside tasks such as feeding cows and shovelling snow, in part because he was bigger (which, in itself, was in part because he was older). Gibbs interview, 14:15.

66 Chanda Pinsent interview, 28:00, 26:00, 40:35.

67 Edell interview, 14:10.

68 Ward, Home Again roundtable, 85:00.

69 Stanley, interview, 26:10.

70 Zimbel, interview, 13:25.

71 Celine Pinsent, interview, 45:30.

72 Home Again roundtable, 12:00, 15:30.

73 Morrison, interview, 20:00.

74 Katz and Katz, interview, 20:10.

75 Ward, interview, 3:00; Home Again roundtable, 85:00.

76 Ward notes that many parents told their children to get an education so they wouldn't have to live like this—"Which was hilarious, because *they* didn't have to do this." Home Again roundtable, 70:50. Parents discuss children's role in their returning to the mainstream throughout the *Back to the Island* interviews. See MacEachern and O'Connor, *Back to the Island*.

77 Edell, interview, 12:30, 1:00. Her family continued to socialize with the back-to-the-land community, however.

78 Arnold, interview, 32:15.

79 Reddin, interview, 1:55, 18:00.

80 Stanley, interview, 2:45.

81 Katz and Katz, interview, 20:10.

82 Rousseau, interview, 20:00.

83 Ward, interview, 48:30.

84 Chanda Pinsent, interview, 37:50.

85 Celine Pinsent, interview, 31:55.

86 E. Gibbs, interview, 4:05.

87 K. Gibbs, interview, 7:35, 19:00.

88 Morrison, interview, 23:55, 21:30. The description of Morrison's mother is from Ward, Home Again roundtable, 25:00.

Index

Note: Page numbers in bold refer to photographs or illustrations.

294   CANADIAN COUNTERCULTURES | COLIN COATES

Warner, Jim, 61, 73
Waste Diversion Ontario, 103
waste management, 19, 70, 169. *See also* recycling
water supply concerns
    Denman Island, 30–31, 33, 40–41, 44, 49
    West Kootenays, 86–90, 91, 92
Watson, Norman, 188
Weaver, Sharon
    biographical highlights, x
    chapter by, 13, 29–54, 58
Weldwood of Canada, 29, 33, 39–41, 48–50
Wells, Andy, 160, 167
Wendt, Pan, 268, 269
West Kootenays. *See also* Slocan Valley Community Forest Management Project (SVCFMP)
    environmental activism in, 85–90, 91–97, 97–98
    immigration and settlement of, 80–81, 83–84
    planning and consultation, 73–74
    uranium mining potential, 79, 85
Wheeler, Doug
    comments about, 189, 197n33
    comments by, 183, 187, 194, 198n35
Whole Earth Catalog, 10, 13, 157, 158, 243
wildlife access and protection, 58, 70–71, 208–9
Williams, Bob, 62–63, 71
Willis, Nancy, 165
Wilson, Jeremy, 62, 71, 73, 97
Winner, Langdon, 155
women. *See also* homebirths
    birth quilts, 246–47
    Blessing Way ritual, 247–49
    home as central to, 235–36
    newspaper contributions of, 34–35
    work done by, 18, 236, 241–42, 273–74

Woodhouse, Keith M., 9
Work, Marvin, 14
work cultures. *See also* work on the land
    Is Five Foundation, 106–7, 120–21
    New Alchemy Institute, 156–57
work on the land
    government funded, 15, 190–93
    as healthful, 1, 7, 11
    nature of, 4, 10–11, 12–13, 263, 265–66
    supplemented by employment, 59
    by women, 236, 241–42, 273–74

Y

Yates, Barbara, 188–89, 193
YCS. *See* Yukon Conservation Society (YCS)
Yorkville neighbourhood, 19–20, 104
youth alienation and rebellion, 15, 20, 180, 188
Yukon
    conservation support, 215
    countercultures of, 201–2, 211, 220–22, 223
    economic development, 205–10, 226n27
    highways, 224n4
    land-use concerns, 213–14
    political culture, 216, 223
    society, 210
Yukon Conservation Society (YCS), 210, 213–16, 222–23, 226n25
Yukon First Nations, 218–20. *See also* Tr'ondëk Hwëch'in First Nation
Yukon Resource Council, 215

Z

Zelko, Frank, 4, 82
Zimbel, Matt, 262–63, 274
Ziner, Feenie, 5, 18

www.ingramcontent.com/pod-product-compliance
Lightning Source LLC
Chambersburg PA
CBHW050807270326
41926CB00026B/4599